# Electronic
# Fault
# Diagnosis

## FOURTH EDITION

### George Loveday C.Eng., M.I.E.E.

Former Head of Division for Electronic Engineering
Bromley College of Higher and Further Education

 LONGMAN

**Pearson Education Limited**
Edinburgh Gate, Harlow
Essex CM20 2JE, England
*and Associated Companies throughout the world.*

First published in Great Britain by Pitman Publishing Limited 1977
Second Edition 1982
Third edition published by Longman Scientific & Technical 1988
Fourth edition 1994
Reprinted 1996, 1999, 2000, 2001

**British Library Cataloguing in Publication Data**
A catalogue entry for this title is available from the British Library

ISBN 0-582-22911-1

Set by 4 in 10/12 Times
Printed in Malaysia, PA

# Contents

# Preface

Preface to the fourth edition

The ability to rapidly diagnose the causes of faults in electronic equipment and circuits is one of the important skills that can be acquired by the electronic technician or mechanic. This book is intended to serve as an introduction to the subject. Naturally, fault diagnosis skill is not achieved easily, since it combines a good understanding of component and circuit operation together with knowledge on testing methods and on how components fail. The exercises throughout the book are designed to assist the student in improving his fault diagnosis technique. The text concentrates mainly on component faults occurring in particular types of circuit rather than on the fault finding techniques used for localising faults in complete electronic instruments or systems. There is, however, a section that deals briefly with system fault finding methods.

The majority of the circuits have been built, tested, and then measurements made under fault conditions, and it is intended that the student should construct or breadboard the circuits as practical project work. For this reason, readily available components, wherever possible, have been chosen.

Primarily the book is intended for students studying City and Guilds 224 Electronics Servicing, and the BTEC Certificate and Diploma courses. However, it is hoped that the text and exercises will prove interesting and helpful to a much wider readership.

**Note to second edition**

The revision has involved the correction of a number of remaining errors, some minor improvements throughout, and, chiefly, an almost complete rewrite of Chapter 8 into an expanded version containing four new exercises.

# Preface to the third edition

The demand for staff with fault finding skills continues to rise and now the Service Engineer is also expected to diagnose and repair faults in a wide range of systems that are microprocessor based. In this edition I have therefore included a basic introduction to fault finding on microprocessor-based systems with an example, using readily available chips, designed around an 8-bit microprocessor.

This expanded version of the book also reflects other important trends in Electronics and includes notes and exercises on opto-isolators, timer ICs, SMPU designs and powerFETs.

# Preface to the fourth edition

In this new edition I have added some new exercises, in particular ones involving the use of ICs, and a new chapter on fault location methods. The latter is needed because many courses now include a fault finding and measurement practical test which has to be passed in order for competence in that area of work to be claimed. This new chapter builds on the basics of fault finding in circuits, as explained in earlier parts of the book, to give the reader experience in the complete fault finding process. The principles of narrowing down a fault to component level from a given system circuit diagram is explained and several small electronic systems are introduced as examples.

# 1 Basics of Fault Diagnosis

## 1.1. Circuits and Test Readings

An electronic circuit is a collection of components connected together to perform a particular electronic function. Each component has its part to play in the operation of the circuit. If any component should fail, then the operation will be drastically changed. For an example, consider the simple relay amplifier circuit of Fig. 1.1. If $R_1$ were to go open circuit, there would be no forward bias current for $Tr_1$. The collector voltage of $Tr_1$ would rise, $Tr_2$ would then conduct, and the relay coil would be permanently energized.

A faulty component produces a certain set of symptoms, which can be used to indicate the component and its type of fault. Such symptoms are, for example, the voltage levels at various points in the circuit.

The voltages,* measured with a standard multi-range meter at the test points of Fig. 1.1, when the circuit is working correctly, and with no input applied, are

| Test point | 1 | 2 | 3 |
|---|---|---|---|
| Voltage | +0·7 | +0·1 | +24 |

However, with $R_1$ open circuit the readings would change to

| Test point | 1 | 2 | 3 |
|---|---|---|---|
| Voltage | 0 | +0·7 | +0·15 |

These readings indicate that $Tr_1$ is non-conducting and, since the base voltage of $Tr_1$ is at zero volts, that possibly $R_1$ is open circuit and cannot supply base current to $Tr_1$. It is worth noting here that a base-emitter short circuit on $Tr_1$ would also cause the same symptoms. A resistance check would be necessary to discover which of the two components

*Throughout the book, the voltage readings in these test point tables are given in volts.

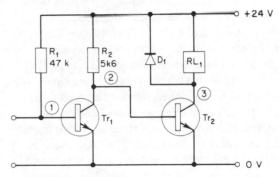

Fig. 1.1 Relay amplifier

was faulty. Component testing is described later in this chapter.

For more complex circuits, especially those which use direct coupling, the effects of one component fault can be extensive. However the fault symptom invariably indicates which component is at fault, and the exercises in the following chapters are intended to give the reader experience in diagnosing faulty components from a given set of symptoms.

Skilful fault diagnosis requires both theoretical knowledge and practical experience. Before attempting the diagnosis of faulty components the technician will need to understand the purpose of the circuit and its operation. This clearly presupposes that he also understands the principle of operation of the various electronic components used. A review of the common components follows in the next section.

## 1.2 Components and Common Faults

Before considering the individual types of component let us look at the ways in which a component can fail. A component can be said to have failed if any one of its constants is out of its specified limit.

For example, if a 5k6 ohm ± 5% resistor actually has a resistance value of 6 k$\Omega$, or if the leakage current of a 64 $\mu$F 12 V electrolytic capacitor is 150 $\mu$A when it is specified as a maximum of 10 $\mu$A, then both components have failed.

1

Both these examples can be described as **partial failures**, since they do not necessarily lead to a complete loss of performance, rather to a small change. Partial failures are especially important when the component is used in a critical circuit position.

The faults we are more concerned with are called **catastrophic failures**, when the failure of the component is both sudden and complete. For example, a resistor goes very high in value, or becomes open circuited, or a diode develops a short circuit between anode and cathode. Such failures lead to a complete loss of performance and are usually accompanied by a drastic change in d.c. bias levels.

As a general rule certain types of component fail in a particular way. When resistors, especially the film type, fail they often go open circuit, since a small break in the resistance spiral is a much more likely event than a short circuit across the whole resistor. Electrolytic capacitors on the other hand are more prone to fail short circuit. Here we are discussing the *way* in which components fail; this should not be confused with the *rate* at which they fail. The reliability of present-day components is extremely high. In other words, the failure rate is low. Resistors, in particular, are very reliable.

Table 1.1 indicates the more probable types of failure for various types of electronic component.

TABLE 1.1

| Component | Common type of fault |
|---|---|
| Resistors | High in value or open circuit. |
| Variable resistors | Open circuit or intermittent contact resulting from mechanical wear. |
| Capacitors | Open or short circuit. |
| Inductors (including transformers) | Open circuit. Shorted turns. Short circuit coil to frame (iron cored types). |
| Thermionic valves | Filament open circuit. Short circuited electrodes (i.e. grid to cathode). Low emission. |
| Semiconductor devices Diodes, Transistors, FETs, SCRs etc. | Open or short circuit at any junction. |

It is perhaps easy to understand failures caused by defects and overloads, but why should a component fail in normal use? Basically the component is ageing because of the stresses that are acting continuously upon it. These stresses are of two kinds, operating and environmental. The **operating stress** is due to the design conditions, and the life of a component can be prolonged by operating it well within its rated maximum value of current, voltage, and power. This is called derating. **Environmental stresses** are those caused by the surrounding conditions. High temperature, high humidity, mechanical shock and vibration, high or low pressure, and corrosive chemicals or dust in the air, are the major adverse conditions. All of these stresses affect the component and cause some deviation from the specification, and finally the component will fail. For example, consider a component subjected to continual cycles of heating and cooling; this may cause the materials from which the component is made to become brittle, and any mechanical shock may then cause the component to fail open circuit.

The effects of adverse environmental conditions can usually be minimized by careful design, and this is increasingly important when an electronic instrument forms an integral part of some industrial manufacturing process where high temperatures, vibration and other hazards are present.

Another cause of component failure is high voltage pulses or "spikes", generated from switched inductive loads, being transmitted along the mains and appearing on internal supply wires. These "spikes" can easily lead to the breakdown of junctions in semiconductor devices.

## 1.3 Operating Principles of Common Active Components

The following is intended only as a review; other texts should be consulted.

### (1) *Semiconductor diodes*

These devices (Fig. 1.2) have a low slope resistance when the anode is positive with respect to the cathode, a typical value being $25\ \Omega$ at a forward current of 1 mA. When the anode is negative with respect to the cathode, the resistance is very high, greater than $100\ M\Omega$ for a silicon diode.

In order to pass current in the forward direction, a small forward bias is necessary, about 200 mV for a germanium device and 600 mV for a silicon diode. The characteristics for small signal diodes are shown in Fig. 1.2. Note that, if a large enough reverse voltage is applied, the diode will break down and will pass a damaging overload current unless limited by some external resistance in series with the diode. Very intense electric fields are set up in the junction when the reverse breakdown voltage is applied and electrons are accelerated to such velocities that they dislodge other fixed electrons from the crystal lattice. These electrons in turn dislodge others, so the process is rapid and cumulative. The effect, put to good use in regulator diodes, is called avalanche. Naturally, a normal diode has a maximum reverse voltage rating that should not be exceeded, usually between 100 and 800 V.

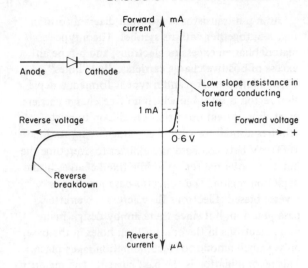

Fig. 1.2 Semiconductor diode

### (2) Voltage Regulator Diodes

The symbol and characteristics of a typical regulator diode are shown in Fig. 1.3. They utilize the zener, or avalanche effect. The device has a higher impurity doping content than an ordinary diode, and this results in a relatively small depletion layer. This means that high field strengths (up to $10^7$ volt/cm) exist in the depletion layer for only small reverse voltages. By controlling the doping, diodes of varying reverse breakdown voltages can be produced – typically from 3·3 V up to 150 V, with power ratings from 250 mW up to and above 75 W.

The device behaves like a normal diode in the forward direction, but in the reverse direction has a very high resistance until the breakdown voltage is reached. At this point the resistance falls to only a few ohms.

A simple application of the voltage regulator diode is shown in Fig. 1.4. The voltage across the diode remains almost constant even if the load current and supply voltage are varied over wide ranges.

### (3) Bipolar Transistors

The transistor is best considered as a device in which a current flowing between the collector and emitter is controlled by a much smaller current flowing between the base and emitter.

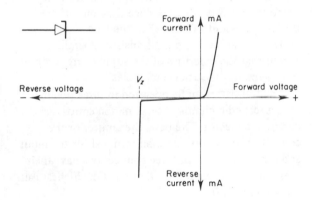

Fig. 1.3 Voltage regulator diode

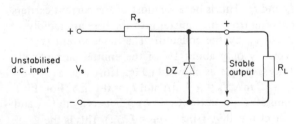

Fig. 1.4 Application of voltage regulator diode

An n-p-n transistor is shown in diagram form in Fig. 1.5, together with its symbol. The n-type material has an excess of electrons, and p-type an excess of positive charge carriers called "holes". When a junction of p and n type is formed, a depletion region is set up in which no free charge carriers exist. For correct operation the silicon transistor requires a small forward bias voltage of approximately +600 mV between base and emitter to overcome the junction potential set up by the fixed charges in the depletion region. The collector base junction is reverse biased. Electrons flow across the emitter base junction, but since there are by design many more electrons in the emitter than holes in the base, only a small amount of recombination takes place. This recombination is the base current. The majority of the electrons diffuse, or spread, across the base until they reach the depletion region of the collector base junction. There they are swept up and collected by the positive field. The electrons which make up the current flowing from collector to emitter are called the majority carriers since they outnumber the small amount of "holes" in the base.

The operation of a p-n-p transistor is similiar except that the polarities of the supplies are reversed and the majority carriers are "holes".

The transistor can be operated in three possible modes, called common base, common emitter, and common collector. The base, the emitter or the collector is made the common terminal for the input and output signals. All three connections have their uses, but the configuration that gives the highest gain is common emitter.

The relationship between the three currents flowing in a bipolar transistor, neglecting any leakage current, can be written as

$$I_e = I_c + I_b.$$

The base current $I_b$ is much smaller than both $I_e$ and $I_c$. This is because most of the current carriers crossing from the emitter into the base are rapidly swept up by the collector. The base current is typically only about 1% of the emitter current. Take example as shown in Fig. 1.6A where $I_e = 10$ mA, $I_c = 9.9$ mA and $I_b = 0.1$ mA. For this transistor, the current gain $h_{FB}$ between emitter and collector is 0.99 (since $h_{FB} = I_c/I_e$). This is the

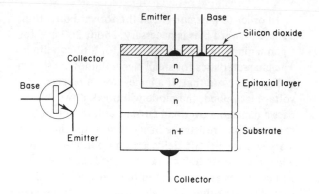

Fig. 1.5 Construction of n-p-n transistor

current gain when the transistor is connected in common base mode.

The current gain $h_{FE}$ between base and collector is 99 (since $h_{FE} = I_c/I_b$). This is the current gain when the transistor is connected in common emitter. It is important to realize that $h_{FE}$ is a parameter that varies widely for transistors of the same type. A glance at any data sheet will quickly show this. A typical spread in $h_{FE}$ may be from 50 to 500. Any bias circuit has to be designed to overcome this large spread.

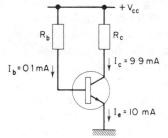

Fig. 1.6A Currents in a simple transistor circuit

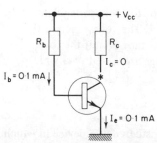

Fig. 1.6B Circuit as 1.6A but with the collector open circuit

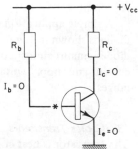

Fig. 1.6C Circuit as 1.6A but with base open circuit

If the transistor fails collector open circuit, there will still be a base current flowing, and this current flows in the emitter (Fig. 1.6B). However, if the base or emitter fails open circuit, apart from a small leakage current, the effective current flow is zero (Fig. 1.6C).

Further work on transistor faults in a single stage common emitter amplifier is dealt with in Chapter 2.

The DARLINGTON is a connection method using two transistors to give high current gain (see Fig. 1.7A). The collectors of the two transistors are connected together and the emitter of the first transistor drives the base of the second. In this way the current flowing in the collector and emitter leads of the output transistor can be controlled by a tiny change in the input current to the base of the first transistor. The overall current gain between $I_{C_2}$ and $I_{B_1}$ can be as high as 2000 and is the product of the current gains of the two transistors.

$$\text{Darlington current gain} \simeq h_{FE_1} \times h_{FE_2}$$

The Darlington can be constructed using two discrete transistors but is commonly available as a complete unit, see Fig. 1.7B. There also are integrated circuits that contain several Darlingtons in one dual in line package.

Because of their very high gain Darlingtons are used in interfacing, as the output stages in power amplifiers and as control elements in power supply regulators.

Two things need to be noted when fault finding on Darlingtons:

(1) The normal on voltage between base 1 and emitter 2 will be approximately $1 \cdot 2$ volts, not $0 \cdot 6$ V, since there are two base emitter junction potentials to overcome, and

(2) when used as a switch the on state voltage between the output collector and emitter will be about 1 V.

### (4) Unipolar Transistors — the FET Family

The operation of FETs differs from the bipolar transistor in that the current flowing through the FET is controlled by a voltage input. The terminals are called drain, source and gate, and the simplified construction of an n-channel junction FET is shown in Fig. 1.8. This is made from a bar of n-type material to which contacts called drain and source are made at each end. Two p-regions formed into the bar directly opposite each other are connected together and are called the gate.

A current will flow between source and drain when the voltage between drain and source is positive. However this current will fall if the gate voltage is made negative with respect to the source. When the gate is negative, depletion regions are formed, and this reduces the channel width between source and drain, thus the current falls. When the gate is made sufficiently negative, say −3 V, these depletion regions meet and the drain current is cut off.

The important feature of a FET is that the drain current is controlled by the voltage on a reverse biased gate to source p-n junction. This means that it has a very high input impedance. A typical FET amplifier circuit is shown in Fig. 1.9.

With small signals the FET behaves like a linear resistor, which can be switched from a few hundred ohms to several hundred megohms by means of the gate voltage. This can be useful in analogue switching circuits and low drift chopper circuits.

Another type of FET is the metal oxide silicon field effect transistor (MOS-FET) sometimes referred

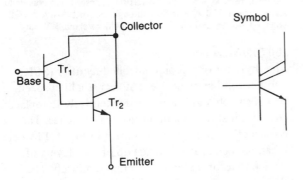

Symbol

Collector

Base

Tr₁

Tr₂

Emitter

**Fig. 1.7A** Basic Darlington

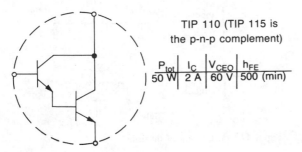

**Fig. 1.7B** Typical discrete unit

TIP 110 (TIP 115 is the p-n-p complement)

| $P_{tot}$ | $I_C$ | $V_{CEO}$ | $h_{FE}$ |
|---|---|---|---|
| 50 W | 2 A | 60 V | 500 (min) |

to as an IGFET or MOST. This device is different in construction to the junction FET, in that the gate is actually separated from the conducting channel by a metal oxide insulating layer. The current flowing through the channel is controlled by the electrostatic field between the gate and the substrate. Such devices have an extremely high input impedance, and care

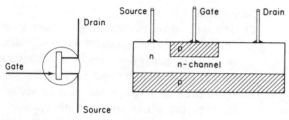

**Fig. 1.8** An n-channel junction FET

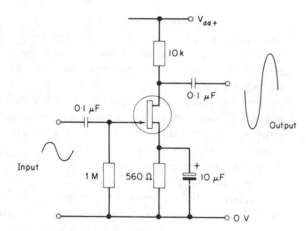

**Fig. 1.9** Typical junction FET amplifier

must be taken in handling and soldering them as stray electrostatic fields can easily damage the thin insulating layer.

The enhancement mode MOSFET is used extensively in the fabrication of digital integrated circuits, i.e. in CMOS logic (CMOS stands for complementary MOSFET logic since it is constructed of p and n channel devices), microprocessors and memory chips. The importance of the MOSFET operation has also led to the development of powerMOSFETs. These are discrete devices, with names such as VMOS, HEXFET or TMOS which are used in many switching and power amplifier applications. The powerFET has a very high input impedance, since it is a voltage operated device and has the

capability of passing very large drain currents in the on-state and of withstanding high off-state voltages. Devices are available that can pass 20 amps, switched by a few volts between gate and source and with off-state voltages between drain and source of 500 volts or more. See Fig. 1.10 for an example.

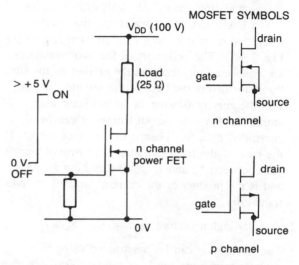

**Fig. 1.10** The use of a powerFET as a simple switch. (When the input rises to +5 V the powerFET is forced into conduction and a current of 4 A will flow through the load. While 'on' the voltage between drain and source should be about 1 V only.)

### (5) *Thyristors and Triacs*

The thyristor or silicon controlled rectifier (SCR) is another solid state device that acts as a high-speed power switch and is now used extensively to replace conventional relays and mechanical switches. The construction and symbol is shown in Fig. 1.11 where it can be seen that it is made up of four layers of semiconductor material in a pnpn sandwich. The thyristor can be made to act either as an open circuit

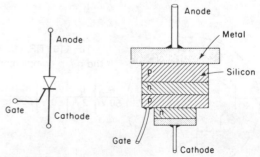

**Fig. 1.11** Construction of thyristor

or as a rectifier depending upon how its gate is used.

Conduction between anode and cathode is blocked in both the reverse and forward direction. The gate has no control over the reverse characteristics, but can be used to switch conduction in the forward direction. When a small signal is applied between gate and cathode the thyristor is turned on, and a large forward current can flow with only a small voltage drop across the device. Once on, the thyristor can only be switched off by reducing the current through it to a value less than what is called the holding current. The holding current is the specified minimum current to ensure continued conduction, and this is usually a few per cent of the maximum forward current. In a.c. power control circuits the thyristor naturally turns off every half-cycle when the supply reverses.

The thyristor can be switched into forward conduction by two other means: (*a*) by exceeding the forward break-over voltage and (*b*) by applying a fast rising voltage waveform between anode and cathode, typically greater than 50 volt/microsec, but normally it is the gate signal that is used to control the point of switch on.

The Triac (Fig. 1.12) is essentially two thyristors connected in reverse parallel. This very useful device can be switched into forward or reverse conduction by means of a control signal applied to its gate. It finds it main application in full wave a.c. power control circuits. (See also Chapter 7.)

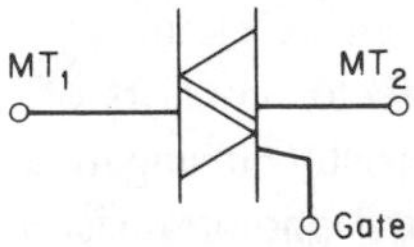

**Fig. 1.12** Triac symbol 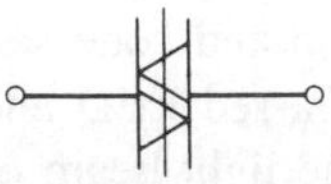   **Fig. 1.13** Diac symbol

## (6) *The Diac*

This component is often used as a triggering device for circuits using thyristors and triacs. Its symbol is shown in Fig. 1.13. The diac will not conduct in either the forward or the reverse direction until a certain threshold voltage is exceeded, usually about 30 V. Once the threshold voltage is exceeded, the diode exhibits a negative resistance as the current increases, while the voltage across the diac falls. In other words above a certain voltage the diac passes a pulse of current.

Since the diac is symmetrical in operation it is extremely useful in making economical trigger circuits for triac full wave a.c. controllers. Some devices are produced which incorporate a diac and triac in one encapsulation; these are called Quadracs.

## (7) *Opto-devices*

Opto-electronics is a term used to describe a wide range of components that link light (optics) with electronics. These include:

    Light emitting diodes (LED)
    LED displays (7-segment, bar graph, dot matrix)
    Photodiodes and Phototransistors
    Opto-couplers and opto-isolators
and Sensors (such as the reflective and slotted opto-switches).

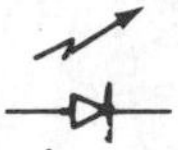

**Fig. 1.14A** LED symbol

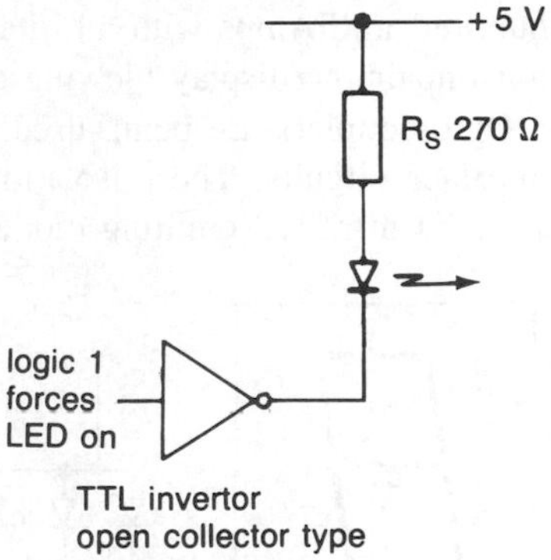

**Fig. 1.14B** Typical use as an indicator

The LED is a semiconductor diode (see Fig. 1.14) that emits light when it is forward biased. Typically the forward bias is 2 V and the current to get a useful light output is between 10 mA to 20 mA. Red, green and yellow diodes are available with higher voltages required by the green (2·2 V) and yellow (2·4 V) types. It is important to limit the current otherwise the diode will burn out and usually a series resistor is included in the circuit. For a 5 V supply this is in the range 150 Ω to 390 Ω.

$$Rs = \frac{(V_S - V_F)}{I_F}$$

Single LEDs make useful on/off or fault indicators and are used extensively on panels and on pcbs.

One of the main points to watch when fault finding is not to apply too large a reverse voltage across the diode — most have a maximum $V_R$ rating of 5 V.

Groups of LEDs are arranged in packages to create bar graph, 7-segment or other display features. The commonly used 7-segment format is shown in Fig. 1.15. The 7 diodes (8 if a decimal point is used) usually have either all cathodes or all anodes commoned up. A common anode connection is shown. Where several 7 segment displays are used for a panel indicator (a digital multimeter for instance) the diodes are multiplexed, that is, switched at a relatively high frequency so that only a selected few are on at any one time. This prevents excessive current being drawn from the supply, which could be several hundred milliAmps with all diodes on, but does not impair the display viewing quality.

Opto-couplers are being used increasingly in interface circuits. These isolating devices consist of an LED infra-red emitting diode in the same package as a light sensitive transistor or switching device. The input and output circuits are electrically isolated from each other since the signal connection is via the optical link between the LED and the sensor. A simple example is shown in Fig. 1.16 for an interface between a microprocessor system and a load in the a.c. mains. When the bit output from the microprocessor system port goes high to logic 1 the LED is forced to conduct, sending a light beam across the gap to a photosensitive triac inside the MOC3030. This triac fires as the mains voltage increases and forces the external main triac to also conduct. Power is applied to the load for as long as the logic signal from the microprocessor system remains at 1.

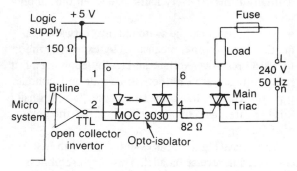

**Fig. 1.16**

Opto-devices are also used in many sensing applications. One very common device, which is available in several different forms, is the reflective opto-detector. It makes an excellent position and code sensor. The detector consists of an infra-red LED and a phototransistor arranged so that the light beam only reaches the phototransistor when it is reflected from a suitable object. The object being sensed must be highly reflective and be positioned a few millimetres from the face of the device. Even then the efficiency is quite low, typically only 2%.

A simple example using such an opto-sensor is given in Fig. 1.17. $R_1$ sets the infra-red LED current to about 25 mA. The current is given by:

$$I_F = \frac{V_{CC} - V_F}{R_1}$$

Where $V_F$, the LED forward voltage, is approximately 2 V.

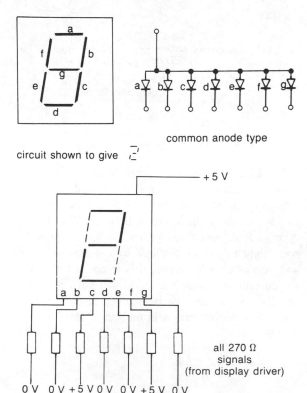

**Fig. 1.15** Seven-segment display format

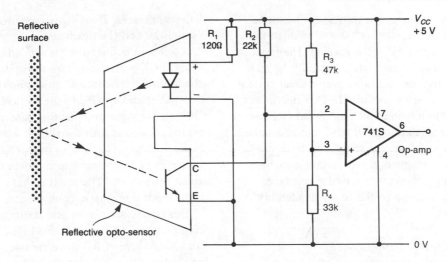

**Fig. 1.17**  Using a reflective opto-detector in a simple circuit

If the light beam is reflected and then picked up by the phototransistor a current of about 0·5 mA will flow through $R_2$ taking the inverting input of the op-amp lower in voltage than the reference level set by $R_3$ and $R_4$

$$V_{ref} = V_{CC} \frac{R_4}{R_3 + R_4} = 2 \cdot 1 \text{ V}$$

The op-amp, wired as a comparator, produces a well-defined pulse when a reflective surface is detected.

### 1.4 Measuring Instruments and Testing Methods

#### (1) *Meters*
To get information about the symptoms of a particular fault, a set of voltage readings at critical points in the circuit must be taken. This information, together with additional information on the circuit performance (i.e. distorted output, overheating component) is usually all that is necessary for correct fault diagnosis. So the only essential piece of test equipment for fault finding is a good, general purpose multi-range meter. This should have a resistance on d.c. ranges of at least 20 kΩ per volt. It is important that the meter has a relatively high resistance, otherwise the loading effect of the voltmeter could lead to incorrect conclusions. Also when measuring

voltages in circuits that have fairly high resistances, the loading effect must be considered.

Take for example the potential divider shown in Fig. 1.18. The voltage across $R_2$ should be 13·3 V. If a meter of 20 kΩ on the 10 V d.c. range is connected across $R_2$ it will actually indicate nearly 10 V. If a higher range of the meter is selected, the meter current is reduced, and a more accurate indication is given. It is always wise to select the highest possible range when measuring voltages in high resistance circuits.

An alternative to the moving coil meter is the small, portable digital multimeter. This displays the

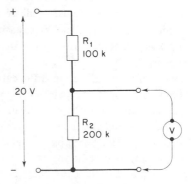

**Fig. 1.18** A voltmeter of 20 kΩ/V measuring the voltage output of a potential divider formed by two resistors of relatively high value. Voltmeter indicates approximately 10 V on 10 V range whereas the true output voltage is 13.3 V

measured voltage, current or resistance on a three or more in-line digital display. The more digits used, the greater the accuracy of the reading. The input resistance of these instruments is typically 10 MΩ, which means that the unit takes only a small current from the circuit being measured. This instrument has replaced the moving coil type meter because of its accuracy, readability and high input resistance. However, unless otherwise stated, in the exercises in the following chapters the measurements have been made using a standard moving coil type meter, mainly because of the ready availability of this instrument.

### (2) *The Cathode Ray Oscilloscope*

Among other useful instruments the next important from the point of view of fault finding is the cathode ray oscilloscope (CRO). This is perhaps the most versatile measuring instrument available. With it, it is possible to measure d.c. and a.c., voltage, current, phase-angle, and a whole range of other quantities. The accuracy depends to a great extent upon the care paid to the instruments' calibration, and in most modern oscilloscopes signals for calibration are built in. The typical input impedance of a CRO is 1 MΩ which has a capacitance of about 20 pF in parallel with it. The input impedance can always be increased by using a special probe unit. A probe is simply a test lead which contains either a passive or an active network at its end or at some point along the lead. The straightforward voltage-divider probe is a basic attenuator with good frequency compensation. The latter is usually adjustable and should be checked before use. A disadvantage of this arrangement is that the signal attenuation is high typically 10:1 or 100:1; this is why the probes are called ×10 or ×100.

The "heart" of an oscilloscope is the cathode ray tube (CRT). This consists of an electron gun, a deflection system and a fluorescent screen. A high velocity, finely focussed electron beam is produced by the electron gun. This beam passes between two sets of plates arranged at right angles. Voltages applied to these plates deflect the beam both horizontally and vertically. The beam finally strikes the screen and a fine point of light is produced. This spot of light can be moved to any part of the screen by applying signals to the horizontal and vertical

deflection plates. These signals are produced from the Y-amplifier and the timebase.

A signal to be measured is applied to the Y-input of the CRO, is attenuated by the switched attenuator (the Y-amplitude control), then amplified by the Y-amplifier and applied to the vertical plates of the CRT. At the same time the timebase unit is triggered to produce a sawtooth signal that, when applied to the horizontal plates, causes the spot to move across the screen at a uniform rate and then fly back and repeat the process. The result is that a bright trace of the input signal appears on the screen.

This trace can only be held stationary if the trigger control on the CRO timebase is correctly set. For a single beam CRO there are two possible triggering modes: external or internal. The external position should be selected only when a trigger signal is available; this feature can be extremely useful when measuring the time or phase relationship between two signals, as will be seen later. The normal mode for the trigger is to select internal. To hold the trace, switch the trigger select switch to INT and then adjust TRIG. LEVEL (or TRIG. STABILITY) until the trace locks.

Suppose we wish to measure the frequency and amplitude of an unknown sine wave signal. The CRO is set up with no input so that first of all the trace is located (some instruments incorporate a beam finder for this purpose). The BRILL and FOCUS controls should be set to give a clear fine line on the screen. The signal to be measured is applied to the Y-input as shown in Fig. 1.19 and the Y-amplitude control and the TIME switch set until the signal can be easily measured. In the example the Y-amplitude control is at 2 V/cm and the time switch is at 0·1 ms/cm. Therefore the unknown signal has an amplitude of 5 V peak and a periodic time of 0·20 ms. The frequency is then given by

$$\frac{1}{T} = \frac{1}{0.2 \times 10^{-3}} = 5 \text{ kHz}$$

As stated previously the CRO is a highly versatile instrument but *always make sure* that it is calibrated correctly and set to a calibrated position.

Many modern CROs have double beams which can be used to display two time-related signals. An example is given in Fig. 1.20 where the signals from an astable oscillator are shown. Only one channel

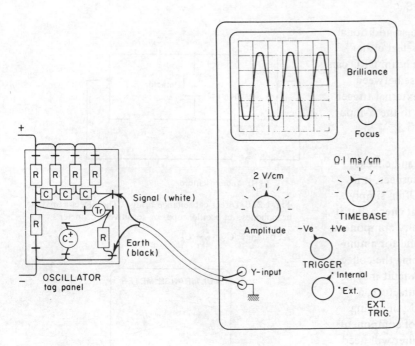

**Fig. 1.19** CRO used to measure a sine wave signal from an oscillator

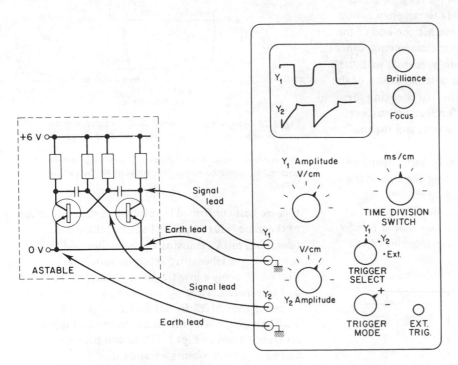

**Fig. 1.20** Double beam oscilloscope used to measure two time-related wave forms from an astable multivibrator. The $Y_1$ beam is triggered on its positive edge

can be used to trigger the Timebase so an additional switch is included that allows one to select either $Y_1$ or $Y_2$ for internal trigger. A single beam CRO can also be used for measurements of phase between signals by applying one signal to the external trigger of the timebase while the other is fed to the Y-input.

### (3) *Simple Component Testing*

When an instrument is being serviced and checks indicate that a certain component is suspect, it is then necessary to confirm the fault. Often simply replacing the component is a sufficient check, but it is always good practice to test the faulty component to verify the type of fault. This is useful for a number of reasons, the most important being the collection of data on component failures. A fault may be caused by defects in component manufacture, a design error, poor production methods, or ageing. Thus, for example, if a large number of components are failing open circuit, the manufacturer will need to be informed so that future defects can be avoided.

Tests to confirm open or short circuit conditions can easily be made using the ohms range of a multi-range meter, but while checking for an open circuit it is usually wise to unsolder and lift one end of the component before making the measurements, otherwise other components that are in parallel with the suspect component will give a false indication of the resistance. An alternative method of checking for an open circuit resistor is to "bridge" the suspect component with a known good one, and then re-check the circuit conditions.

"Leaky" capacitors can also be tested using an ohmmeter, again by disconnecting one end of the capacitor from the circuit. A good electrolytic should indicate a low resistance initially as the capacitor charges, but the resistance should rapidly increase to approach infinity. Open circuit capaci-

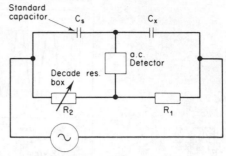

**Fig. 1.21B** Direct capacitance bridge. The detector may be headphones, an oscilloscope, or sensitive a.c. meter

At balance: $C_x = \dfrac{R_2}{R_1} C_s$

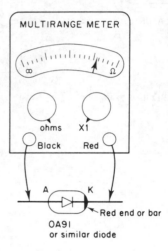

**Fig. 1.22** Using a semiconductor diode to determine the polarity of a multirange meter when switched to the ohms range.
The meter measures a low resistance, indicating that the black terminal is connected to the positive plate of the internal battery.

tors are best confirmed by placing another capacitor of the same value in parallel and checking circuit operation, or by removing the capacitor and testing it on a simple laboratory set-up as shown in Fig. 1.21A using a low frequency generator at 1 kHz and two meters. Here $C_x = I/2\pi f V_0$ with an accuracy of better than ±10% for values from 1000 pF to 1 $\mu$F. An even better method is to use a simple a.c. bridge as shown in Fig. 1.21B to compare the unknown capacitor against a standard.

Tests on diodes, transistors and other semiconductor devices can also be made using the ohms range of a multimeter.

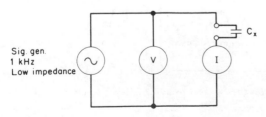

**Fig. 1.21A** Simple laboratory set-up to measure capacitance

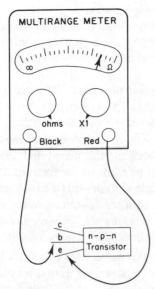

**Fig. 1.23A** Measuring the junction resistance of an n-p-n transistor with a multirange meter. Forward bias on base emitter. A low resistance (typically less than 1 kΩ) should be indicated

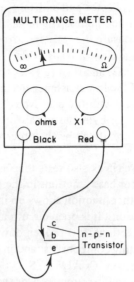

**Fig. 1.23C** Reverse bias on emitter base. A high resistance (greater than 100 kΩ) should be indicated

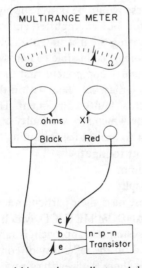

**Fig. 1.23B** Forward bias on base collector. A low resistance (less than 1 kΩ) should be indicated

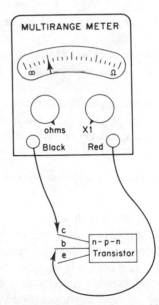

**Fig. 1.23D** Reverse bias on collector base. A high resistance (greater than 100 kΩ) should be indicated

First it is necessary to determine how the internal battery in your multimeter is connected. For example, in one typical instrument the common terminal (marked black) has a positive voltage on the resistance range. If you do not know the connections for the particular meter you are using, the polarity can be determined by connecting the multimeter (on ohms range) to an electronic voltmeter, or by measuring the forward and reverse resistance of a semiconductor diode of known polarity. See Fig. 1.22.

Having established the ohmmeter lead polarity, you can discover a great deal about a transistor. First identify the device leads if not known (see Fig. 1.23). Measure the forward and the reverse resistance between pairs of leads until you find two that measure high (over 100 kΩ) in both directions. These must be the collector and emitter (provided the transistor is a good one). The remaining lead is the base. Now measure the resistance from base to one of the other transistor leads; it should be low in one direction (1 kΩ) and high (greater than 100 kΩ) in the other. If the low resistance occurs when the ohmmeter lead with the positive voltage is connected to the transistor base, the transistor will be n-p-n type. Of course, it will be the other way round for p-n-p.

The above check also tests that both emitter base and collector base junctions in the transistor are good. If either junction shows up high resistance in both directions, it is open circuit; and low resistance in both directions, it is broken down.

When testing components, and in particular transistors, FETs and ICs ALWAYS

(1) Check for power supplies near the actual components, and in the case of ICs directly on the appropriate pins.

(2) Do not use large test probes because they can easily cause shorts.

(3) Avoid the use of excessive heat when unsoldering a component and do not unsolder with the unit switched on.

(4) Never remove or plug in a device without first switching off the power supply. Components can be damaged easily by the excessive current surges.

### 1.5 Fault Finding on Electronic Instruments and Systems

The previous sections dealt with component failures and simple component testing, and the following chapters are concerned with actual component faults in one portion of a complete instrument, in other words for example in the power supply or the oscillator section of an instrument. However, when a complete instrument is returned for repair, the service engineer must first locate which block of the instrument is faulty before he can locate the actual component that has failed. There are various methods used to narrow down the fault to one block, but before these are discussed, it is useful to consider some fairly obvious but often overlooked points:

(1) The service engineer must have a MAINTENANCE MANUAL with up-to-date circuit diagrams. This manual should also give the figures of the per-

formance specification.

(2) The engineer must have all the necessary TEST EQUIPMENT. Usually a list of instruments and any special instructions are in the maintenance manual.

(3) The engineer then has to DEFINE THE FAULT ACCURATELY. This point is most important. It is no good trying to locate a fault that is vaguely defined. The symptoms must be accurately noted and this means that a functional test must be made on the instrument.

For example consider that a signal generator has been returned for repair with a suspect power supply failure. Before taking off the cover and checking power supply lines, the service engineer would

(a) check the mains fuses and if not blown
(b) check for sine wave output on all ranges,
(c) then make notes of the fault symptoms.

The complete circuit of most electronic instruments can be broken down into a series of functional blocks; for example in a general purpose sine wave generator these would be power supply, variable sine wave oscillator, buffer amplifier, and output attenuator. By treating the instrument in blocks, rather than as a whole, it is possible to narrow down the search for a faulty component first of all to one block, then by measurements within that block to locate the actual faulty component. The methods used to decide which *block* is faulty are

(a) Input to output (or beginning to end).
(b) Output to input.
(c) Random.
(d) Half-split.

All of these have their particular advantages and uses. The RANDOM METHOD, which implies a totally non-systematic approach, is rarely used. A method based on the reliability of components can also be used when there is a wealth of service knowledge and experience concerning a particular instrument. For example a service engineer might make the reasonable assumption that, because a particular electrolytic capacitor has been at fault in 60% of the instruments recently returned, it is a strong possibility that the next faulty instrument also has a faulty electrolytic capacitor. He would naturally check this first, and in most cases save valuable service time. It must be stressed, however, that this method depends upon the availability of a large amount of data on the reliability of the various components within an instrument. Most service engineers would use a logical systematic approach to system fault location.

The INPUT TO OUTPUT and OUTPUT TO INPUT methods are examples of this systematic approach. The method is fairly obvious. A suitable input signal (if required) is injected into the input block and then measurements are made sequentially at the output of each block in turn, working either from the input towards the output or from the output back to the input, until the faulty block is located. This logical method is the one most service engineers use on equipment containing a limited number of blocks.

The HALF-SPLIT method is very powerful in locating faults in instruments made up of a large number of blocks in series. Take for example a superhet radio receiver shown in block diagram form in Fig.1.24. Since there are eight blocks it is possible to divide the circuit in half, test that half, decide which half is working correctly, then split the non-functioning section into half again to locate the fault. An example is the best way of really understanding the method. Assume that a fault exists in the demodulator of the receiver, the sequence of tests would be as follows:

(a) Split in half, inject signal into the input of (1) (the aerial circuit), and check output at (4)(IF). Output correct. Therefore the fault is somewhere in blocks (5) to (8).

(b) Split blocks (5) to (8) in half by checking output of (6). Input signal can be left at (1). No output.

(c) Leaving signal at (1), check output from (5). Output should be correct, indicating that the faulty block is (6), the demodulator.

You can try this method for yourself by assuming that the fault is in block (3) for example, and you will find that the number of checks necessary to locate the fault is still three. On average, four tests would be required by using the input to output

technique. The half-split method is most useful when the number of components or blocks in series is very large, for example where several series plug and socket connections are used, or for heater chains in valve equipment. There are, however, several assumptions made for the half-split: (a) that all components are equally reliable; (b) that it is possible and practical to make measurements at the desired point; and (c) that all checks are similar and take the same amount of time. These assumptions will not always be valid and it is up to the service engineer to then decide the best method of approach.

The half-split method can also be easily complicated by

(a) An odd number of series units.

(b) Divergence: an output from one block feeding two or more units.

(c) Convergence: two or more inputs being necessary for the correct operation of one unit.

(d) Feedback: which may be used to modify the characteristics of the unit or in fact be a sustaining network for an oscillator.

When using any of the methods as described, try and use the method, or a combination of them, that will locate the faulty block in a system in the shortest possible time.

I shall end this introductory chapter with a simple exercise that illustrates some of the features required in successful fault diagnosis. This is a circuit, see Fig. 1.25, that acts as a lamp failure indicator. This type of circuit might be used in a car to monitor the operation of a rear brake light. It consists of only a few components and can be

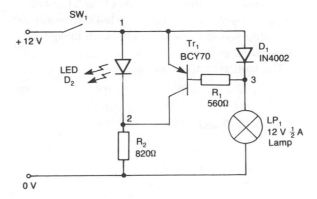

Fig. 1.25  Lamp failure indicator

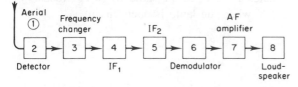

Fig. 1.24  Block diagram of superhet radio

easily built by the reader. A 12 V 0·5 A filament lamp is assumed in the circuit. If a lower wattage lamp has to be used (say 12 V at 100 mA) the power diode $D_1$ should be made up of two small signal diodes in series, this is because the operation of the circuit depends on the increased forward voltage that will appear across a diode when its forward current is relatively high.

Suppose the filament lamp is operating correctly, then the volt drop ($V_F$) across $D_1$ is sufficient to forward bias $Tr_1$. This p-n-p transistor conducts and causes the LED current to be zero. The current through $R_2$ flows through the transistor not the LED, and the LED is off.

If the filament lamp fails, and the failure mode would be open circuit, there will be no current through $D_1$, and $Tr_1$ turns off. The LED now conducts with the 15 mA current through $R_2$ now flowing through the LED.

Now let us consider some basic fault conditions. Suppose the filament lamp and the LED are both on. The faulty component(s) could be:

(a)  $D_1$ short circuit, a possible failure mode for a rectifier diode but rather unlikely.

(b)  $R_1$ open circuit.

or

(c)  $Tr_1$ open circuit in some way. For example either base-emitter open circuit or collector-emitter open circuit.

A few test readings with a voltmeter would then allow us to locate the faulty component.

Suppose $D_1$ had failed short circuit; in this case the voltmeter readings at points (1) and (3) with respect to 0 V will be identical. In other words the volt drop across $D_1$ will be zero even though the filament lamp is on.

The fault symptoms will be:

| Fault | $TP_1$ | $TP_2$ | $TP_3$ |
|---|---|---|---|
| $D_1$ S/C | 12 V | 10·1 V | 12 V |
| Both LED and $LP_1$ on | | | |

If the fault was caused by $R_1$ or $Tr_1$ open circuit the readings would be:

| Fault | $TP_1$ | $TP_2$ | $TP_3$ |
|---|---|---|---|
| $R_1$ O/C | 12 V | 10·1 V | 11·1 V |

You can easily verify these results for yourself.

Now let us consider another fault where the symptom is that the LED fails to operate when the filament lamp is removed. In this case the more probable components failures are:

(a)  The LED open or short circuit.

(b)  $R_2$ open circuit.

or

(c)  $Tr_1$ short circuit between its collector and emitter.

The faulty component can again be located by measurement. In all the following readings $LP_1$ is removed.

| Fault | $TP_1$ | $TP_2$ | $TP_3$ | Symptom |
|---|---|---|---|---|
| LED S/C | 12 V | 12 V | 11·5 V | In all these faults |
| LED O/C | 12 V | 0 V | 11·5 V | the LED fails to |
| $R_2$ O/C | 12 V | 10·6 V | 11·5 V | indicate that $LP_1$ |
| $Tr_1$ S/C C to E | 12 V | 12 V | 11·5 V | is open circuit |

It can be seen from these readings that it is not possible to locate the fault to either the LED short circuit or to $TR_1$ short circuit since the voltage readings are identical. Some additional tests have to be made; for example, the transistor's collector could be disconnected from the circuit. Then, with power reapplied, if the LED is on, the faulty component must be the transistor.

Another point worth noting is that when the circuit path to a component is open circuit the result of reading the voltage with a meter is to pass a small current through the device. This is illustrated in Fig. 1.25 when the resistor $R_2$ limiting the current to the LED is considered open circuit. When the meter is connected, $TP_2$ indicates 10·6 V. In the same way, $TP_3$ reads 11·5 V when the lamp is open circuit.

# 2　Single Stage Transistor Amplifier

## 2.1 Basic Principles

This section is concerned with the effects of individual component failures in a single-stage common emitter amplifier. The circuit shown in Fig. 2.1 usually has eight components. Remembering that resistors can fail high or open circuit; capacitors either open or short circuit; and that the transistor can fail open or short circuit between any of its connections; it can be seen that a total of at least twelve faults is possible. For each of these faults a unique set of conditions will exist.

Before considering any fault conditions, the operation of the circuit must be understood. In a class A amplifier a mean current flows through the transistor, and the input signal causes this current to either increase or decrease. This change in collector current then develops a voltage signal across the collector load resistor $R_3$. The operating point of the collector voltage, that is the d.c. voltage between collector and the 0 V rail, should be a value that allows equal positive and negative swings of the output signal. As a rough approximation, $V_C$ should be half the supply voltage. The whole purpose of the bias components ($R_1$, $R_2$, $R_4$) is to fix this operating point, and to keep it stable. Stability is very important since a number of factors cause the operating point to change. For silicon transistors the most important is the change in current gain $h_{FE}$. This can have a value from 50 to 500 for the same type of transistor and, obviously without some form of stabilizing circuit, the operating point would change drastically in the circuit every time the transistor was changed. The bias circuit achieves stabilization by fixing the value of the base voltage $V_B$ and by keeping it constant irrespective of changes in the base current. To do this the values of $R_1$ and $R_2$ must be chosen so that the current flowing through them is much greater than the transistor base current. These resistors form a potential divider

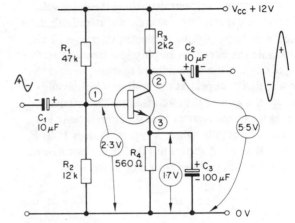

**Fig. 2.1** Single stage class A transistor amplifier; normal working voltages

and if we neglect base current, the d.c. base voltage can be calculated from

$$V_B \simeq \frac{V_{CC}}{(R_1 + R_2)} R_2$$

and the emitter voltage $V_E$ is given by

$$V_E = V_B - V_{BE}$$

where $V_{BE}$ is the forward bias voltage between base and emitter, typically 0·7 V for a silicon transistor. Then

$$\text{Emitter current } I_E = \frac{V_E}{R_4}$$

and as we are neglecting base current, $I_E \simeq I_C$ so the d.c. voltage at the collector $V_C$ is given by

$$V_C = V_{CC} - I_C R_3$$

Now, since $V_B$ is fixed, the d.c. current through the transistor will be fixed, and this gives the operating point $V_C$.

In operation, the circuit is an example of series negative feedback. Imagine that the collector current increases, thus causing the operating point to fall.

17

The emitter current also increases, raising the emitter voltage $V_E$. However, since $V_B$ is fixed by the potential divider, any increase of $V_E$ must reduce the voltage between base and emitter of the transistor, and this in turn causes a decrease of collector current. This tends to counteract the original rise to stabilize the operating point.

Having set the correct bias with the resistors, the a.c. input and output signals must be coupled to and from the circuit without disturbing the d.c. levels. To do this capacitors $C_1$ and $C_2$ are used. Both these should be fairly high-value electrolytics, say $10~\mu F$, to enable the circuit to amplify low frequencies. Capacitor $C_3$, the decoupling capacitor, ensures that no a.c. signals appear at the emitter which would reduce the gain of the circuit. Since the internal resistance of the emitter base junction is quite low, $C_3$ must be high in value. A typical value is $100~\mu F$.

For the circuit shown in Fig. 2.1 the calculations of the d.c. bias voltages would be as follows:

$$V_B = \frac{V_{CC}}{(R_1 + R_2)} \cdot R_2$$

$$= \frac{12}{47~k\Omega + 12~k\Omega} 12~k\Omega = 2 \cdot 4~V$$

$$V_E = V_B - V_{BE} = 2 \cdot 4 - 0 \cdot 7 = 1 \cdot 7~V$$

$$V_C = V_{CC} - I_C R_3$$

where $I_C = I_E = V_E/R_4$

This is assuming that the current gain is high and that the base current can be neglected. This is nearly always the case. Therefore

$$V_C = 12 - (3 \cdot 05~mA \times 2 \cdot 2~k\Omega) = 12 - 6 \cdot 7 = 5 \cdot 3~V$$

Putting these in table form we have, for the calculated values:

| TP | 1 | 2 | 3 |
|----|-----|-----|-----|
| V | 2·4 | 5·3 | 1·7 |

In fact when the circuit is built, the actual voltages measured with a $20~k\Omega/V$ meter will be slightly different. This is to be expected since the bias resistors used have a tolerance of 10%.

The actual readings were

| TP | 1 | 2 | 3 |
|---------|-----|-----|-----|
| Meter reading | 2·3 | 5·5 | 1·7 |

This shows the close agreement between the calculated and the measured values. When fault finding on any circuit always try to make a rough calculation of the voltage you would expect. This can be an invaluable guide as to which parts of the circuit are functioning correctly.

Now let us consider the effect of component failures, taking each in turn.

## 2.2 Resistor Faults

### $R_1$ OPEN CIRCUIT (Fig. 2.2)

| TP | 1 | 2 | 3 | No output |
|----|---|-----|---|-----------|
| MR | 0 | +12 | 0 | signal |

When $R_1$ goes open circuit, the current flowing in $R_2$ and the base is zero. It follows that the transistor is cut off so both the emitter and base voltages are zero. Since no collector current is flowing the voltage dropped across the collector load $R_3$ is zero and the collector voltage itself is the same as the supply voltage $V_{CC}$.

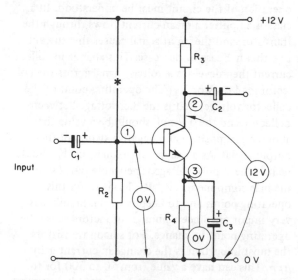

**Fig. 2.2** $R_1$ open circuit

### $R_2$ OPEN CIRCUIT (Fig. 2.3)

| TP | 1 | 2 | 3 | Grossly distorted |
|----|-----|-----|-----|-------------------|
| MR | 3·2 | 2·6 | 2·5 | output; negative going signals clipped. |

Without $R_2$ in circuit the current that was flowing through $R_2$ now tries to flow into the base of the transistor. But the value of base current will be limited by the transistor current gain, so less current flows through $R_1$. This means that the base voltage must rise. The base current in fact rises to a value that completely "turns on", or saturates the transistor, so that the collector voltage is only about 0·1 above the emitter voltage.

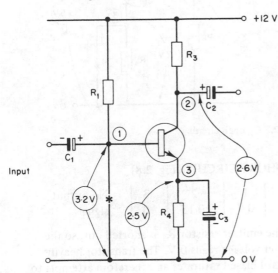

**Fig. 2.3** $R_2$ open circuit

### R₃ OPEN CIRCUIT (Fig. 2.4)

| TP | 1 | 2 | 3 | No output |
|---|---|---|---|---|
| MR | 0·75 | 0·1 | 0·1 | signal |

Without $R_3$ in circuit the collector current is zero, so any current flowing in the emitter must now be supplied from the base. The base/emitter junction acts like a forward biased diode placing $R_4$ in parallel with $R_2$. Since $R_4$ is a low value resistor (560Ω) the emitter voltage falls to a very low value. The base voltage, as expected, is about 650 mV greater than the emitter voltage.

It might be reasonable to assume that the voltage reading at the collector would be zero since the resistance is open circuit. However, when the meter is connected it presents a high resistance path from the collector to 0 V, and the base/collector junction acts like a forward biased diode passing a small current through the meter.

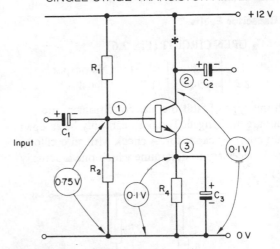

**Fig. 2.4** $R_3$ open circuit

### R₄ OPEN CIRCUIT (Fig. 2.5)

| TP | 1 | 2 | 3 | No output |
|---|---|---|---|---|
| MR | 2·3 | 12 | 2 | signal |

With an open circuit between emitter and 0 V, no currents flow through the transistor. The collector voltage therefore rises to $V_{CC}$. The voltage at the base is fixed by the potential divider $R_1$ and $R_2$ and since the base current is small in comparison to the current through $R_2$ this voltage hardly changes at all.

As with the previous example when the meter is connected between emitter and 0 V, a small emitter current flows so the voltage indicated at the emitter is slightly higher than normal.

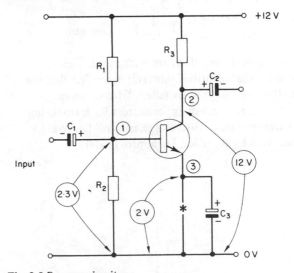

**Fig. 2.5** $R_4$ open circuit

## 2.3 Capacitor Faults

### $C_1$ or $C_2$ OPEN CIRCUIT (Fig. 2.6)

| TP | 1 | 2 | 3 | No output |
|----|-----|-----|-----|-----------|
| MR | 2·3 | 5·5 | 1·7 | signal |

With this type of fault the bias conditions of the circuit are unchanged. The fault can only be an open circuit coupling capacitor. A check with an oscilloscope is necessary to determine which one is actually faulty.

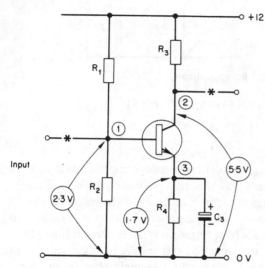

**Fir. 2.6** $C_1$ or $C_2$ open circuit

### $C_3$ OPEN CIRCUIT (Fig. 2.7)

| TP | 1 | 2 | 3 | |
|----|-----|-----|-----|-----------|
| MR | 2·3 | 5·5 | 1·7 | Low gain |

Again the bias conditions are unchanged. The symptom that identifies this fault is the fact that the amplifier voltage gain has fallen. With $C_3$ open circuit, a.c. signals will appear across $R_4$ introducing negative feedback. The voltage gain will fall to a value given by $R_3 \div R_4$, i.e. approximately 4.

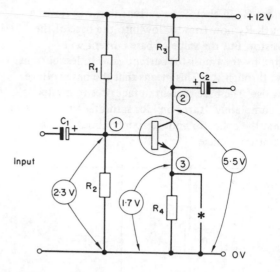

**Fig. 2.7** $C_3$ open circuit

### $C_3$ SHORT CIRCUIT (Fig. 2.8)

| TP | 1 | 2 | 3 | No output |
|----|-----|------|-----|-----------|
| MR | 0·7 | 0·15 | 0 | signal |

The emitter resistor $R_4$ is shorted out, so the emitter voltage reads 0 V. The transistor heavily forward biased saturates and therefore attempts to pass a large current. However, the transistor current is limited to a value given by $V_{CC} \div R_3$ which prevents the transistor from being damaged. The base voltage must be 0·7 V higher than the emitter.

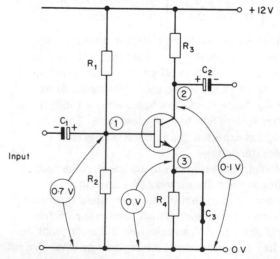

**Fig. 2.8** $C_3$ short circuit

## 2.4 Transistor Faults

### COLLECTOR/BASE JUNCTION OPEN CIRCUIT (Fig. 2.9)

| TP | 1 | 2 | 3 | |
|----|-----|----|-----|-----------|
| MR | 0·75 | 12 | 0·1 | No output |

Since the collector is open there can be no collector current flowing, so the voltage at TP2 rises to +12 V. The base/emitter junction now acts as a forward biased diode in a similar way as for the fault of $R_3$ open circuit.

### COLLECTOR/BASE JUNCTION SHORT CIRCUIT (Fig. 2.10)

| TP | 1 | 2 | 3 |
|----|---|---|-----|
| MR | 3 | 3 | 2·3 |

As with any short circuit a clue to the fault is given by the fact that the voltages on the base and collector are equal. With this fault the circuit effectively reduces to $R_3$ in series with the base/emitter diode and $R_4$. The resistance of this path is much lower than $R_1$ and $R_2$, so the effect of the latter resistors can be neglected. The current flowing in $R_4$ is given by

$$I = \frac{V_{CC} - V_{BE}}{R_3 + R_4} = \frac{12 - 0·7}{2·76 \text{ k}\Omega} = 4 \text{ mA}$$

The voltage at the emitter will then be $I \times R_4 = 2·3$ V. The voltages at TP1 and 2 will be 0·7 V higher than this, sufficient to forward bias the base/emitter diode.

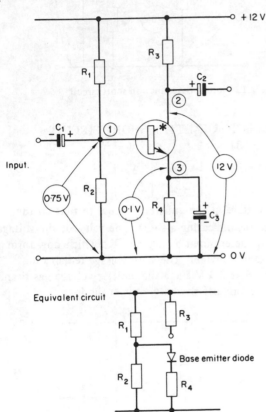

**Fig. 2.9** Collector base junction open circuit

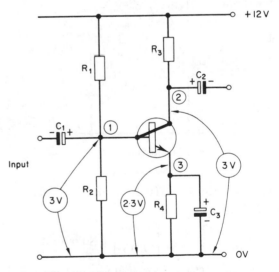

**Fig. 2.10** Collector base junction short circuit

## EMITTER/BASE JUNCTION OPEN CIRCUIT
(Fig. 2.11)

| TP | 1 | 2 | 3 | No output |
|----|---|---|---|-----------|
| MR | 2·3 | 12 | 0 | signal |

With this fault there can be no current flowing in the transistor. The voltage drops across $R_3$ and $R_4$ are zero, so the collector voltage rises to $V_{CC}$ and the emitter voltage is 0 V. The voltage on the base is determined by the potential divider $R_1$ and $R_2$ and therefore remains at 2·3 V. There is no difference in the symptoms if the base or the emitter connection to the junction is open circuit.

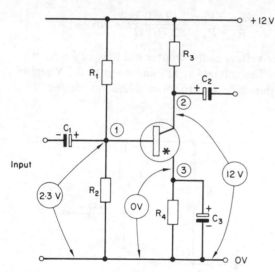

**Fig. 2.11** Emitter base junction open circuit

## EMITTER/BASE JUNCTION SHORT CIRCUIT
(Fig. 2.12)

| TP | 1 | 2 | 3 | No output |
|----|---|---|---|-----------|
| MR | 0·13 | 12 | 0·13 | |

The voltages at TP1 and TP3 will be equal and at a low value since $R_4$, a low resistance, is placed directly in parallel with $R_2$. With a shorted base/emitter junction all transistor action ceases, so the collector voltage rises to $V_{CC}$.

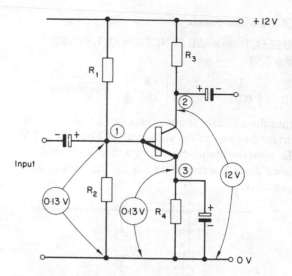

**Fig. 2.12** Emitter base junction short circuit

## COLLECTOR/EMITTER SHORT CIRCUIT
(Fig. 2.13)

| TP | 1 | 2 | 3 |
|----|---|---|---|
| MR | 2·3 | 2·5 | 2·5 |

The voltage at the emitter is equal to that on the collector, indicating a short. The value of the voltage will be determined by $R_3$ and $R_4$ which now form a potential divider. The base voltage remains unchanged at 2·3 V since the emitter voltage has risen, thus cutting off the base/emitter diode.

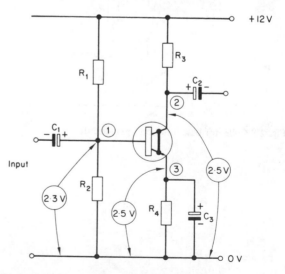

**Fig. 2.13** Collector emitter short circuit

## Questions

(1) Estimate the voltages that would be indicated by a 20 kΩ/V meter between the test points and 0 V in all circuits in Fig. 2.14. In all cases silicon diodes and transistors are used

(2) The circuit for this exercise is a common emitter amplifier (Fig. 2.15). It has a voltage gain of about 80. First calculate the voltages you would expect to measure between the test points and 0 V. Follow the procedure as outlined.

The following table shows the voltage readings at the test points for various component faults. In each case state which component is faulty and the type of fault.

|  | 1 | 2 | 3 | *Additional symptom* |
|---|---|---|---|---|
| Fault A | 0·16 | 9 | 0·16 | No output |
| Fault B | 1·5 | 9 | 1 | No output |
| Fault C | 0·85 | 9 | 0·15 | No output |
| Fault D | 1·5 | 1·45 | 1·45 | No output |
| Fault E | 1·5 | 4·5 | 0·8 | Very low gain |
| Fault F | 0 | 9 | 0 | No output |

(3) The circuit shown in Fig. 2.16 is a common base amplifier. With this type of amplifier the input is applied at the emitter and the output taken from the collector. The bias circuit is identical in operation to the potential divider bias of the common emitter.

Again calculate the voltages you would expect to measure with a 20 kΩ/V meter between the test points and 0 V. Then determine which component or components could cause the following fault conditions.

| Fault | 1 | 2 | 3 |
|---|---|---|---|
| A | 0 | 0 | 12 |
| B | 0 | 3 | 12 |
| C | 3·8 | 3 | 3·8 |
| D | 1·1 | 1·7 | 1·1 |
| E | 5·2 | 5·9 | 5·9 |
| F | 3·7 | 4·4 | 3·8 |

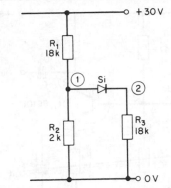

Fig. 2.14A

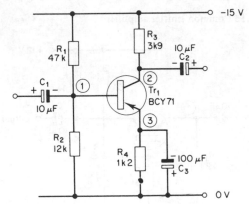

Fig. 2.14B Tr₁ is silicon

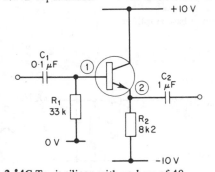

Fig. 2.14C Tr₁ is silicon with an $h_{FE}$ of 40

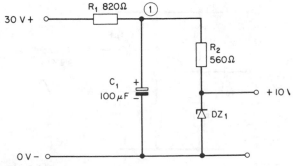

Fig. 2.14D

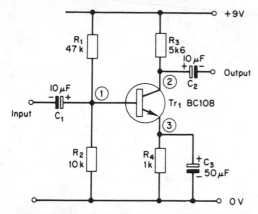

Fig. 2.15 Common emitter amplifier

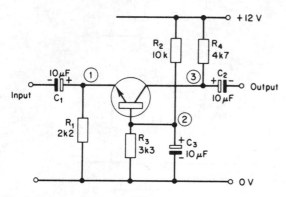

Fig. 2.16 Common base amplifier

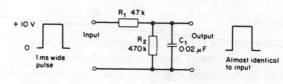

Fig. 2.17C

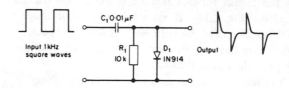

Fig. 2.17D

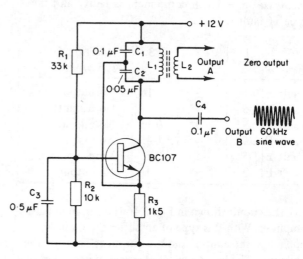

Fig. 2.17E

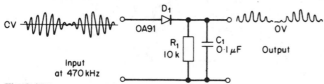

Fig. 2.17A

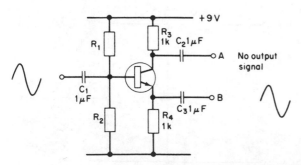

Fig. 2.17B

(4) In the circuit of Fig. 2.17 no d.c. bias voltages are given. Instead the input and output waveforms are displayed. In each case one component can be considered to be faulty. State which component and the type of fault.

# 3   Power Supply Circuits

## 3.1 Basic Principles of D.C. Power Supplies

All electronic instruments require a source of d.c. power before they will operate. Sometimes the source is a battery, but more usually the power is obtained from a unit that converts the normal single phase a.c. mains supply (240 V at 50 Hz) to some different value of d.c. voltage.

The function of the power supply is to provide the necessary d.c. voltage and current, with low levels of a.c. ripple (mains hum) and with good stability and regulation. In other words it must provide a stable d.c. output voltage, irrespective of changes in the mains input voltage and of changes in the load current.

A further important requirement of a modern unit is that it should be able to limit the available output current in the event of an overload (current limiting) and also limit the maximum output voltage. Damage to sensitive components, such as ICs, in the instrument can easily occur if excessive voltages appear on the power supply lines.

There are various methods of achieving a stable d.c. voltage from the a.c. mains, but only two methods are commonly used. These are

   (i)  Using a linear stabiliser
   (ii)  Using a switching mode stabiliser

Both have their advantages and disadvantages as will be seen. The switching mode power supply unit (SMPU) is a relatively new innovation and finds its main use in high-power applications (100 W upwards).

## 3.2 The Linear Stabilized Power Unit

The block diagram of a conventional power unit is shown in Fig. 3.1. The TRANSFORMER serves two main purposes: it isolates the equipment d.c. power lines from the mains supply, and it changes the level of the a.c. mains voltage to some lower or higher value. The ratio of the secondary voltage to primary voltage is determined by the number of turns on each winding.

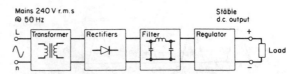

Fig. 3.1 Block diagram of conventional power unit

The RECTIFIER unit converts the a.c. voltage from the transformer secondary winding into pulses of unidirectional current. Three types of rectifier circuit are used for single phase: the half-wave, the full-wave, and the bridge. These, together with their output waveforms, are shown in Fig. 3.2.

The half-wave rectifier, although being a simple circuit, has the main disadvantage of low efficiency. The diode conducts only on one half of the cycle, so the efficiency cannot be greater than 50 per cent. The full-wave rectifier uses two diodes, each conducting on alternate half cycles to give much higher efficiency. However, to achieve this, a transformer with a centre tapped secondary winding is necessary. This means that twice the number of turns is required on the secondary winding. This circuit was common when valve rectifiers were in use, since it was cheaper to wind extra turns on the transformer than to use more valves. The bridge rectifier, now the circuit of choice, uses four diodes to achieve rectification over the whole cycle, and no centre tap is required. The four diodes can now be supplied in one encapsulated unit, which is more convenient and somewhat cheaper than wiring in four separate diodes. However should one part of the encapsulated bridge circuit fail, the whole unit then has to be replaced.

Following the rectifier is the FILTER which serves to smooth out the pulses received from the rectifier. The circuit can have either a capacitive or an inductive input as shown in Fig. 3.3. The inductive filter, or choke input filter, is more commonly used when the power unit has to supply a large load current. On low power equipment a capacitive input filter is more typical. The input capacitor, called the

(i) Half wave

Note that $V_d \triangleq 0.7$ V for a silicon rectifier

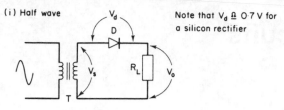

Output waveform assuming 50 Hz a.c. input

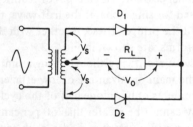

$V_{pk} = \sqrt{2} \, V_s - V_d$

(ii) Full wave rectifier

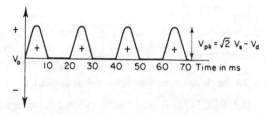

Output waveform assuming 50 Hz a.c. input

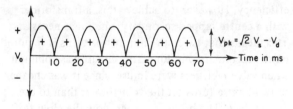

$V_{pk} = \sqrt{2} \, V_s - V_d$

(iii) Bridge rectifier

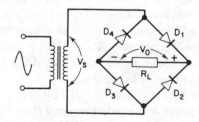

**Fig. 3.2** Single phase rectifier circuits

(a) Capacitor input

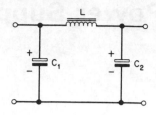

(b) Choke input

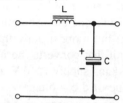

**Fig. 3.3** Filter circuits

"reservoir", is used as a storage device for electric charge. Let us suppose a reservoir capacitor is connected to the output of a half-wave rectifier as in Fig. 3.4. When the diode conducts on the positive half cycle, the capacitor is charged and a large pulse of current is taken. The voltage across the capacitor then rises to nearly the peak value of the a.c. secondary voltage. When the secondary voltage begins to fall,

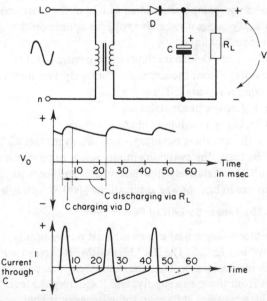

**Fig. 3.4** Reservoir capacitor connected to the output of a half wave rectifier. Showing typical wave forms, assuming 50 Hz a.c. input.

the diode becomes reverse biased, and the capacitor now discharges through the load resistor. The voltage across the load now more nearly represents a d.c. level, but superimposed upon it is an alternating waveform, called the ripple. The value of the ripple amplitude depends upon the size of the capacitor and the load resistance. To achieve low values of ripple a high-value electrolytic capacitor, typically 500 $\mu$F or more. has to be used. A number of points should be noted concerning the reservoir capacitor.

(a) Since it is an electrolytic, it is polarized and must be connected correctly in the circuit.

(b) Its d.c. working voltage must be greater than the peak of the transformer secondary voltage.

(c) It must be physically large since it has to absorb large pulses of current when it is charging, the peak values of which may be several amperes. If too small a capacitor is fitted, it may overheat and possibly explode!

Check the ripple current rating.

The other components of the circuit form a low pass filter, which reduce still further the output ripple voltage. Typical values for the iron-cored inductor are 1 to 5 Henries and for $C_2$, 500 $\mu$F. In some circuits these components may be omitted, especially when an efficient regulator is used. The inductor also is often replaced by a wire-wound resistor of low value, say 22 $\Omega$, in which case there will be a voltage drop across this component resulting in a lower output voltage.

The last block is the REGULATOR, which is used to keep the output voltage constant irrespective of changes in the mains input voltage and of changes in the load current. These two functions are called line stabilization and load regulation respectively. All linear regulators comprise

(a) a control unit

(b) a reference element (usually a zener diode)

(c) an error amplifier

as shown in Fig. 3.5.

In operation, the circuit compares a portion of the d.c. output voltage with the reference voltage. Any difference between the two levels is amplified by the error amplifier, and the output fed to the control unit. The stability and regulation of the output voltage depends upon the stability of the reference element and the gain of the error amplifier. High gain Op-Amps in IC form are now commonly used as the

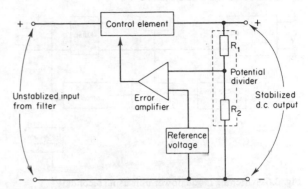

**Fig. 3.5** Basic block diagram of a linear regulator

error amplifier to give power supplies of excellent performance.

The main advantage of the linear regulator is that the output is continuously controlled to give good stabilization against mains input changes and good regulation against load current changes.

A typical specification for an output voltage of +15 V @ 100 mA load current is

| | |
|---|---|
| Line stability | 10 000:1 |
| | (a 10 V change in mains supply giving a 1 mV change in d.c. output) |
| Output ripple | 0·1 mV pk-pk at full load |
| D.C. output impedance | 0·05 ohms |
| Temperature coefficient | 200 $\mu$V per degree C |
| Load regulation | 0·033% from zero to full load (i.e. an output change of 5 mV) |

The limitation in the linear regulator circuit is that good performance is achieved at the expense of inefficiency. Power is dissipated and lost in the series control transistor and this power loss increases with load current. A large heat sink is required to ensure that the junction temperature of the series transistors is kept within its rated value. For power units supplying above about 100 W, the switching mode regulator becomes a preferred alternative.

### 3.3 Switching Mode Power Supplies (SMPU)

There are two main variations of this type. In one, a fast switching transistor is used as the control element in the regulator (Fig. 3.6). This transistor is switched on and off at a frequency above audio (usually 20 kHz). The d.c. output voltage, after being

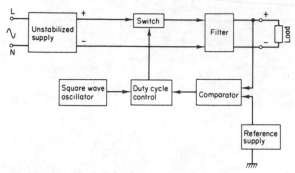

**Fig. 3.6** Switching mode power unit using secondary switching

smoothed by a low pass filter, is controlled by varying the mark-to-space ratio of the switching signal. Such techniques are known as secondary switching. The error signal, generated by comparing the d.c. output with a reference level, is used to control the duty cycle of a free-running oscillator. The advantage of this type of circuit is that the series transistor heat dissipation is greatly reduced, hence greater regulator efficiency.

Another form of the SMPU is shown in Fig. 3.7 and uses a principle called primary switching. The mains supply itself, after rectification and smoothing, is switched at high frequency by high voltage switching transistors. With this method the transformer following the switching transistors can be much smaller than the bulky 50 Hz transformer required in conventional supplies. Regulation is achieved by again varying the switching duty cycle of the transistors. Naturally RF suppression circuits must be

included to reduce the switching spikes that would otherwise be fed back into the mains supply. This SMPU offers considerable advantages in terms of efficiency, reduction in heat loss, and reduction in overall volume. However it does not possess the regulation performance that can be achieved in the linear circuit. Switching mode supplies are now commonly used where large currents at low voltage are required, as in equipment using many digital ICs.

### 3.4 Power Supply Protection Circuits

Some form of protection must be incorporated in even the simplest power supply. A common form is the standard fuse which serves to disconnect the unit from the mains supply when an overload or short occurs. A power unit may have fuses in the line and neutral mains wires, and also a fuse in the d.c. unstabilized line. Fuses usually do not blow soon enough to protect the series transistor in the regulator if the output is shorted, and so some form of current limiting device is used. A simple circuit for achieving this is shown in Fig. 3.8 where the load current flows through a low value current monitoring resistor. If the load current increases beyond a predetermined value, the voltage developed across this resistor turns on $Tr_2$ which in turn tends to turn off $Tr_1$, the series transistor.

Over-voltage protections can be provided by a circuit which senses the d.c. output voltage, and compares it with a reference level as in Fig. 3.9. If the d.c. output voltage rises above $V_z$ a signal is generated which triggers the thyristor and this short circuits the output, either blowing a d.c. line fuse or operating the current limit. Such circuits are called "crowbars". Naturally the fault must be cleared before the circuit can be reset.

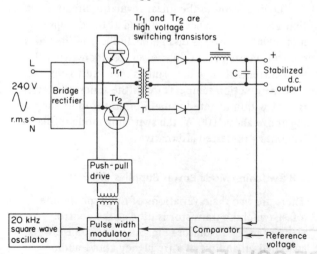

**Fig. 3.7** Switching mode regulator using primary switching

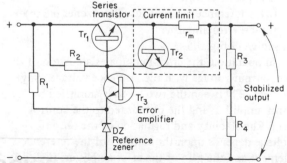

**Fig. 3.8** Linear regulator with simple current limiting circuit

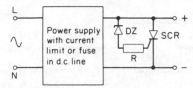

Fig. 3.9 Over-voltage protection circuit (crowbar)

### 3.5 Testing Power Supply Circuits

The main parameters which ought to be measured either in a test department or by the service technician after he has repaired a power unit are the following:

(a) D.C. output voltage

(b) Available d.c. output current

(c) Output ripple voltage at full load

(d) Stabilization against mains supply changes

(e) Regulation from zero to full load

These can all be measured using a standard test set-up as shown in Fig. 3.10.

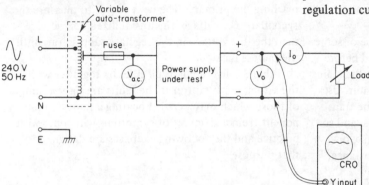

Fig. 3.10 Laboratory test set-up for measuring the performance of a power supply unit

The d.c. output voltage should be measured, and if necessary adjusted, when the unit is fully loaded. However it is sometimes advisable to measure the output on a low load and then gradually increase the load current to maximum. There should, of course, be little change in the output voltage.

The peak-to-peak ripple amplitude can be checked best by measuring at the output with an oscilloscope. A sensitive a.c. range must be selected because the ripple should be quite low, typically less than 20 mV.

Measurement of stabilization and regulation requires that any small change in d.c. output be carefully noted, and therefore a digital voltmeter is often necessary. For stabilization measurement, the unit should be fully loaded and the change in d.c.

output voltage noted for say a ±10% change in the a.c. input. The mains input can be varied using an adjustable auto-transformer as shown. Then if, for example, the d.c. output changed by 50 mV from 10 V, i.e. an output change of 0·5%, then the line stabilization would be 40:1.

Load regulation is measured, keeping the a.c. input constant, by noting the change in output when the load is varied from zero to full load.

$$\text{Load regulation} = \frac{\text{Change in d.c. output}}{\text{D.C. output on no load}} \times 100\%$$

For example suppose the output changed by 20 mV from 10 V. The load regulation is

$$\frac{20 \times 10^{-3}}{10} \times 100\% = 0.2\%$$

To obtain fuller information on a power supply's performance it is often necessary to plot the load regulation curve. This is a plot of output voltage against load current. A typical result for a unit with current limiting is shown in Fig. 3.11.

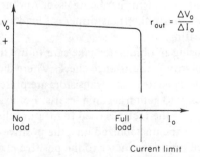

Fig. 3.11 Typical load regulation plot for a power unit with a current limit.
Between no load and full load current the change of output voltage should be very small.

### 3.6 Fault Finding Techniques and Typical Fault Conditions

When a faulty power unit is returned for repair, the fault has to be isolated to some particular portion of the unit. The fault may lie in the transformer, the rectifier, the filter section, or the regulator, and measurement with a voltmeter will be necessary to locate the fault.

However its probably best to start diagnosis with a few rather obvious but often overlooked checks.

First measure the d.c. output voltage. If this is zero, the next check should be on the mains input. Is the mains supply reaching the transformer primary? If it isn't, there is the possibility of a faulty plug (relatively simple to repair!), open circuit mains wires, or a blown fuse. If the fuse is suspected, always test its continuity with an ohmmeter, never rely on just a visual inspection. It is also worth noting that both the live and neutral wires may have a fuse in circuit, so make sure both are checked.

If the fuse is blown it has done so because of some fault condition and the fault must be cleared before a new fuse is fitted. Resistance checks (with the mains unplugged!) must be used to locate such a fault. Use an ohmmeter to measure the resistance of the transformer primary, the secondary, the rectifiers, and so on. The winding resistance depends, of course, on the size of the transformer. The primary resistance, for a medium size transformer, should be low, typically about 50 Ω. The secondary, usually supplying a lower voltage, may have a resistance of only a few ohms. Detecting shorted turns on a winding can therefore be quite difficult. Wherever possible compare the measured resistance with any available data on the type of transformer being used. Another useful check is to run the transformer off load and test for overheating.

When using an ohmmeter take care to use the correct polarity for resistance checks where diodes, electrolytic capacitors and transistors are present. It is all too easy to get misleading results. For example, in Fig. 3.12, if the meter is used to measure the resistance of the unstabilized line, the positive prod (connected inside the meter to the positive plate of a battery) should be placed on the positive line and the negative prod to earth. If the meter is reversed

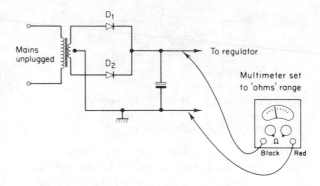

**Fig. 3.12** Using an ohmmeter to measure the resistance across the unstabilized line

there will be a low resistance path through the rectifiers and a leakage path through the capacitor.

Returning to the faulty power unit, suppose however that the fuse is intact, and that the mains is reaching the primary. The next step is to measure the secondary a.c. voltage, the unstabilized d.c. voltage then the d.c. voltage in the regulator and so on, until the fault is located.

Table 3.1 lists some typical faults together with the associated symptoms. The faults are only a sample of those which may occur. Locating a faulty component from a given set of symptoms will come with practice and the following exercises are designed for that purpose.

TABLE 3.1. Typical Faults on Power Supply Units

| FAULT | SYMPTOMS |
|---|---|
| Mains transformer, open circuit primary or secondary | D.C. output zero. Secondary a.c. zero, high resistance primary or secondary. |
| Mains transformer, shorted turns on primary or secondary | Two possibilities: (a) mains fuses blown or (b) low d.c. output and transformer overheating because of excessive current being drawn. |
| Mains transformer, windings shorting to frame or screen | Fuses blown. Low resistance between windings and earth. |
| One diode in bridge open circuit | Circuit behaves as a half-wave rectifier. Lower d.c. output with poor regulation. Increased ripple at 50 Hz not 100 Hz as should be the case. |
| One diode in bridge short circuit | Mains fuse blown, since secondary winding will be practically shorted every other half cycle. A resistance check across each arm of bridge is required, measuring the resistance of each diode in the forward and reverse direction. |
| Reservoir capacitor open circuit | Low d.c. output with very high values of a.c. ripple on output. |
| Reservoir capacitor short circuit | Fuses blown. D.C. resistance of unstabilized line low in both directions. |
| Error amplifier in regulator open circuit | High d.c. output that is unregulated. No control signal for the series element. |
| Series transistor open circuit base emitter | Zero d.c. output. The unstabilized d.c. will be slightly higher than normal since no current is being drawn. |
| Reference zener short circuit | Low d.c. output. Possibility of series transistor overheating. |

### 3.7 Exercise: Power Unit with a Simple Linear Regulator (Fig. 3.13)

This unit incorporates most of the features discussed earlier and is designed to give an output of 12 V at 100 mA. The output resistance is less than $0.5 \, \Omega$, the load regulation better than 0.5%, and the ripple less than 5 mV peak to peak on full load.

The unstabilized d.c. is obtained from a bridge rectifier circuit and a reservoir capacitor of 3300 $\mu$F. The transformer has a secondary voltage of 12 V r.m.s. so the unstabilized voltage across $C_1$ will be approximately $12\sqrt{2}$ V, i.e. about 16 V.

The reference voltage is provided by a 5.6 V zener diode; a 400 mW device such as a BZY88 C5V6 is ideal. $Tr_1$ is the d.c. error amplifier which compares a portion of the d.c. output voltage, the voltage across $R_4$, with the reference. Any difference between the two voltages is amplified by $Tr_1$ and the amplified signal is fed to the base of $Tr_2$. Consider for example the case when the d.c. output falls when more load current is taken; the base voltage of $Tr_1$ decreases and $Tr_1$ conducts less current. Therefore $Tr_1$ collector voltage rises, and this rise in voltage is coupled through $Tr_2$, which acts as an emitter follower to counteract the original fall in output. Thus the circuit operates to maintain the output as nearly constant as possible.

Since the output load current is only 100 mA, then $Tr_2$ (BFY51) need not be mounted on a heat sink. The circuit is not provided with a current limit, so should an excessive current be drawn, say by shorting the output, then $Tr_2$ would undoubtedly burn out. A current limit can be added if required as shown in Fig. 3.8, but for the purposes of the exercise we shall assume no current limit.

The normal d.c. voltages measured with a standard multirange meter are as follows:

| Test point | 1 | 2 | 3 | 4 |
|---|---|---|---|---|
| Voltage | 16 | 13 | 5.8 | 12.2 |

First let's consider the following fault condition:

| TP | 1 | 2 | 3 | 4 |
|---|---|---|---|---|
| V | 17.5 | 17.5 | 0 | 0 |

The d.c. output is zero, but the unstabilized input to the regulator has risen, indicating that little current

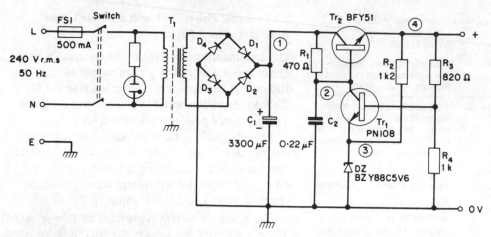

**Fig. 3.13** A 12 V 100 mA power supply
Transformer: primary 0–240 V a.c.
secondary 12 V 250 mA
R.S. type 196–303 or similar
Rectifiers IN4001

is being drawn. Also TP2 is at the same voltage as
TP1. This further shows that no current at all is
flowing through $R_1$ into the base of $Tr_2$. The only
possible fault is that $Tr_2$ has an open circuit base
emitter junction. Note that if $R_1$ were open circuit
TP2 would be at zero volts.

Consider a fault condition when all test points are
at zero volts. Further inspection shows that the fuse
has blown. Resistance checks give the primary resistance as 43Ω, the secondary as 4Ω, but TP1 to 0 V
is zero ohms. The fault in this case can only be $C_1$
short circuit.

**Questions**  The following table lists a series of
fault conditions; in each case state which component
or components could cause the fault and give a
supporting reason.

| | 1 | 2 | 3 | 4 | Other symptoms |
|---|---|---|---|---|---|
| Fault | | | | | |
| A | 16 | 15 | 14·5 | 14·5 | |
| B | 11 | 6 | 4·8 | 5 | Increased ripple |
| C | 16 | 15 | 5·8 | 14·5 | Poor regulation |
| D | 17·5 | 0 | 0 | 0 | |
| E | 16·5 | 2·1 | 0 | 1·5 | |
| F | 17·5 | 17·5 | 0 | 0 | |
| G | 16 | 7·5 | 5·8 | 7 | |
| H | 16 | 5·9 | 5·9 | 5·2 | Poor regulation |

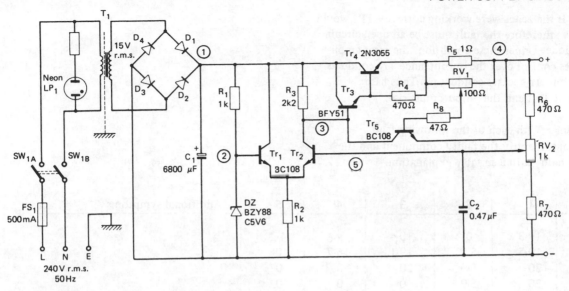

**Fig. 3.14** 10 V 1 A stabilized power supply with current limit
Rectifiers IN5401

### 3.8 Exercise: Stabilized Power Supply with Current Limit Circuit (Fig. 3.14)

This unit is designed to the following specification:
D.C. output: 10 to 15 V adjustable, at 1 A.
Current limit: adjustable from 500 mA.
Output ripple: 20 mV pk-pk on full load.
Load regulation: better than 1%.
Output resistance: 0·1 Ω

In this circuit the error amplifier is formed by $Tr_1$ and $Tr_2$ which are wired as a differential amplifier. This gives improved stabilization of output voltage against changes in ambient temperature. The reference element, the 5·6 V zener is connected to $Tr_1$ base, while a portion of the output is applied to $Tr_2$ base, Any difference between these two levels is amplified and fed to the base of $Tr_3$ and so to $Tr_4$ to control the output level. $Tr_3$ and $Tr_4$ are wired as a darlington amplifier to give high gain round the circuit. To maintain a safe junction temperature for $Tr_4$, at ambient temperatures up to 50°C, this transistor should be mounted on a heat sink which has a thermal resistance of not greater than 10°C per watt.

The current limiting is provided by the circuit comprising $R_5$, $RV_1$, $R_8$ and $Tr_5$. A portion of the voltage developed across $R_5$, caused by the load current passing through it, is applied via $R_8$ (a base

is sufficiently high to cause $Tr_5$ base voltage to exceed about 600 mV, then $Tr_5$ begins to conduct and it draws current away from $Tr_3$ base, thus causing the series element to limit its conduction. The output can be short circuited and the load current will be limited to about 1 A. The actual value of load current limiting is determined by the setting of $RV_1$.

An over-voltage crowbar protection circuit can be included as shown in Fig. 3.9. If the output voltage exceeds the value of zener DZ the thyristor is triggered into conduction and this short circuits the output voltage bringing the current limit into operation.

The normal voltages measured at various test points are as follows ($RV_2$ is adjusted to give an output of 10 V).

| TP | 1 | 2 | 3 | 4 | 5 |
|---|---|---|---|---|---|
| V | 19·5 | 5·9 | 11·9 | 10 | 5·9 |

Suppose we had the following fault conditions:

| TP | 1 | 2 | 3 | 4 | 5 |
|---|---|---|---|---|---|
| V | 19·2 | 19·1 | 19·1 | 16·7 | 9·2 |

TP1 has fallen slightly, indicating that more current is taken, while all other voltages have risen much

higher. If the zener were working correctly TP2 would
be 5·9 V; therefore the fault must be an open circuit
zener diode  Under these conditions the error amplifier loses control over the output since $Tr_1$ conducts
heavily and $Tr_2$ ceases conducting. Thus base
voltage of $Tr_3$, and the output, must rise.

**Questions**   With each of the following fault
conditions, identify the faulty component and the
type of fault, with a suitable explanation.

| Fault | 1 | 2 | 3 | 4 | 5 | Additional symptoms |
|---|---|---|---|---|---|---|
| A | 19·3 | 0 | 10·4 | 8·6 | 4·2 | |
| B | 19·5 | 5·9 | 19·1 | 16·7 | 0 | |
| C | 20 | 5·9 | 20 | 0 | 0 | |
| D | 20 | 5·9 | 0 | 0 | 0 | |
| E | 19·5 | 5·9 | 11·9 | 10 | 5·9 | Current limit will not operate |
| F | 19·5 | 5·9 | 18·9 | 17·4 | 9·1 | Poor regulation |
| G | 19·3 | 5·9 | 6·1 | 18·6 | 6 | Poor regulation |
| H | 0 | 0 | 0 | 0 | 0 | $LP_1$ glowing |
| I | 0 | 0 | 0 | 0 | 0 | $FS_1$ fuse blown. Resistance of primary 48 $\Omega$. Resistance of secondary 6 $\Omega$. Resistance TP1 to ground + ve prod on TP1 6800 $\Omega$ |

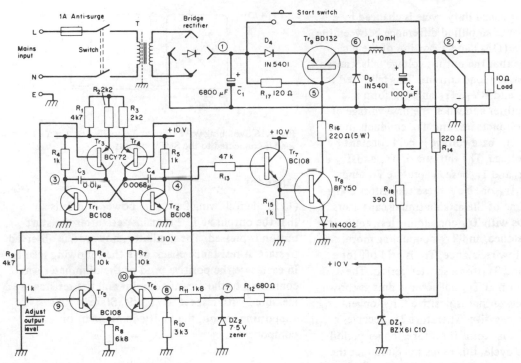

**Fig. 3.15** Switching mode regulator, 20 V at 2 A
Transformer: primary 0–240 V
secondary 25 V r.m.s. @ 3 A
Bridge rectifier: IN5401

### 3.9 Exercise: Switching mode power supply (Fig. 3.15)

This unit, designed to give a stable d.c. output of 20 V at 2·5 A uses secondary switching. This mode of operation was chosen for two reasons: firstly the components required are fairly easy to obtain and relatively cheap and secondly the resulting circuit is not too complex. Primary switching regulators, although being superior, require relatively expensive high voltage switching transistors, and a specially wound high frequency transformer.

In this design the switching transistor is a BD132; this, although being designated an audio transistor, has an $f_T$ of 60 MHz and can pass a maximum current of 3 A. If a higher output current is required a transistor such as a TIP 2955 could be used; this has a maximum collector current of 15 A. The regulator is designed to give an introduction to the techniques involved in switched power units. The operation can best be understood by referring to Fig. 3.15. The various operations are performed by the following transistors and components:

| | |
|---|---|
| Square wave oscillator | $Tr_1$ and $Tr_2$ freq. approx. 5 kHz |
| Duty cycle control | $Tr_3$ and $Tr_4$ |
| Comparator | $Tr_5$ and $Tr_6$ |
| Reference | $DZ_2$ a BCY88 7·5 V zener |
| Switch drive | $Tr_7$ and $Tr_8$ |
| Series switch | $Tr_9$ |

An unstabilized d.c. voltage of approximately 35 V is developed across $C_1$ by the bridge rectifier. The transformer has therefore a secondary voltage of 25 V r.m.s. and a rating of about 100 VA.

The unstabilized voltage is switched at a frequency of about 5 kHz by $Tr_9$. The switching signal is supplied via $Tr_8$, $Tr_7$ from the astable multivibrator formed by $Tr_1$ and $Tr_2$. The duty cycle, or mark/space ratio, of this astable is controlled by the conduction of $Tr_3$ and $Tr_4$. A relatively low frequency is chosen mainly for demonstration of operation, since the signal is just audible when the circuit is operating. Decreasing $C_3$ and $C_4$ will increase the frequency to above the audio range.

The switching signal duty cycle is changed by a d.c. level that is the amplified difference between the reference voltage ($DZ_2$) and a portion of the d.c. output. Imagine that the output voltage falls, caused say by an increase in load current. This fall is fed to $Tr_5$ base via $R_9$ and $RV_1$. $Tr_5$ and $Tr_6$ form a differential amplifier and when the base voltage of $Tr_5$ falls, $Tr_5$ conducts less and $Tr_6$ conducts more. This is because $Tr_6$ base voltage is held constant by the reference voltage. The outputs of $Tr_5$ and $Tr_6$ are connected to $Tr_3$ and $Tr_4$ base. Therefore $Tr_3$ tends to turn off and $Tr_4$ on. Now these transistors control the discharge time of the astable timing capacitors $C_3$ and $C_4$. Thus with $Tr_3$ conducting less, equivalent to a higher resistance, and $Tr_4$ conducting more, equivalent to a low resistance, $Tr_2$ is held off for a longer period, and $Tr_2$ for a shorter period. The collector waveform at $Tr_2$ collector is thus as shown in Fig. 3.16, a signal that is positive for a longer period than it is negative. Therefore $Tr_9$ receives a switching signal that turns it on for a longer period of the switching cycle; this tends to counteract the original fall in output level.

Since the series transistor is switched on and off continuously its collector dissipation is quite low and only a small heat sink say 5 cm² is required. The switched output from $Tr_9$ is filtered by $L_1C_2$. During the OFF period for $Tr_9$, $D_4$ and $R_{17}$ minimize the collector leakage current, and $D_5$ transfers the stored energy in $L_1$ to the load. Therefore both $D_4$ and $D_5$ must have a peak current rating equal to the maximum collected current of $Tr_9$.

In order to start the circuit a switch is included so that power can be supplied from the unstabilized output to the multivibrator, however once the circuit is running the stabilized d.c. output is used to power the multivibrator.

The normal working voltages for a load of 2 A are as follows:

| TP | 1 | 2 | 3 | 4 | 5 |
|----|----|----|-----|-----|-----|
| MR | 35 | 20 | OSC | OSC | OSC |

| | 6 | 7 | 8 | 9 | 10 |
|--|-----|-----|-----|---|-----|
| | OSC | 7·2 | 5·1 | 5 | 7·8 |

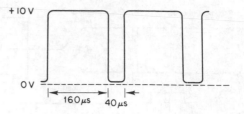

Fig. 3.16 Switching wave form at $Tr_2$ collector when a heavy load is connected to the SMPU output

## Questions

(1) In the following faults the power unit fails so that the output at TP2 remains zero after the start button is pressed. The oscillator, however, is observed to start. A resistance check gave the following results. In each case the positive prod of the ohmmeter was connected to the measured point and the resistance measured with respect to chassis. State, with a supporting reason, the faulty component or components.

| | 1 | 2 | 5 | 6 | Trans. prim. | Trans. sec. |
|---------|------|------|--------|--------|------|----|
| Res. | | | | | | |
| A | 7 kΩ | 10 Ω | 160 kΩ | 12 Ω | 33 Ω | 2 Ω |
| B | 7 kΩ | 10 Ω | 7 kΩ | 500 kΩ | 33 Ω | 2 Ω |
| C | 7 kΩ | 10 Ω | 7 kΩ | 12 Ω | 33 Ω | 2 Ω |

(2) Sketch the time related waveforms you would expect to measure using a double beam CRO at $Tr_2$ and $Tr_8$ collectors, for the conditions of low output load current. Assume that the CRO is triggered from the $Tr_2$ waveform on the $Y_1$ beam.

(3) State the symptoms for the following faults:

(a) $C_1$ short circuit.
(b) $Tr_8$ base emitter short circuit.
(c) $Tr_4$ base emitter open circuit.
(d) $R_{16}$ open circuit.
(e) $R_{14}$ open circuit.

(4) The power unit fails so that the fuse is blown. A d.c. resistance check gave the following results. State, with a supporting reason, the probable fault.

| TP | 1 | 2 | 5 | 6 | Trans. prim. | Trans. sec. |
|------|------|------|------|------|------|----|
| Res. | 7 kΩ | 10 Ω | 7 kΩ | 12 Ω | 33 Ω | 2 Ω |

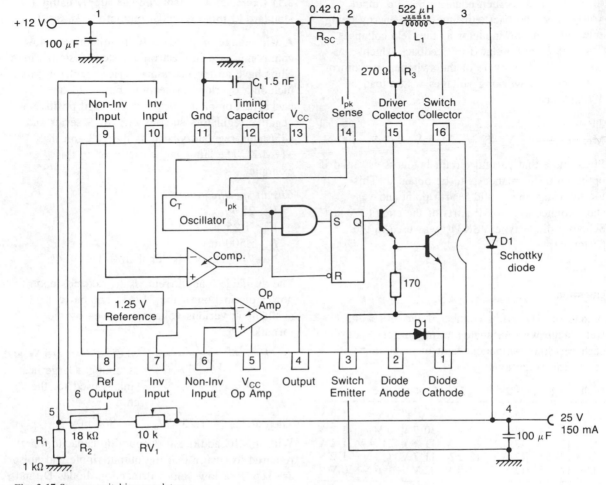

**Fig. 3.17** Step-up switching regulator

### 3.10 Exercise: step-up switching regulator using a standard IC (Fig. 3.17)

The principle of secondary switching is used in several standard switching regulator ICs. The 78S40 chip is a typical example; it contains a stable 1·25 V reference, a controlled duty-cycle oscillator, an active current limit, a comparator, a series Darlington switch capable of passing 1·5 A and withstanding 40 V, and a power diode. With a few external components this chip can be used to create step-up, step-down and inverting regulators. The external components are a timing capacitor to fix the oscillator frequency, a short circuit current sense resistor, feedback resistors to set the voltage,

and LC filter components. The specification for the step-up regulator is:

$V_{in}$: 12 V
$V_{out}$: 25 V nominal
$I_{out}$: 150 mA max
$I_{sc}$: 165 mA
$f_0$: 15 kHz

The step-up arrangement has the switch (the transistors inside the 78S40) connected in parallel after the inductor. This ensures that the output voltage is always *higher* than the input voltage. When the switch closes, current flows through the inductor and energy is stored in the inductor's magnetic field. When the switch opens the voltage across the load is then equal to the d.c. input voltage *plus* the back emf due to the energy being released from the inductor as the field collapses. The voltage is regulated by feedback which controls the duty cycle of the switching waveform.

The normal working voltages with a load of 100 mA are

| TP | 1 | 2 | 3 | 4 | 5 | 6 |
|---|---|---|---|---|---|---|
| MR | 12 V | 12 V | 12 V | 24·5 V | 1·2 V | 1·2 V |

Please note that careful circuit layout is essential if optimum performance is to be obtained. This means using one "solid" earth point and very short connections to all parts of the circuit. A fast Schottky diode type VSK340 was used in the prototype.

## Question

A load of 100 mA is connected and all readings were taken with an analogue multimeter. State in each case the component fault or faults that will set up the conditions.

| Fault | 1 | 2 | 3 | 4 | 5 | 6 |
|---|---|---|---|---|---|---|
| A | 12·2 V | 12·2 V | 0 V | 0 V | 0 V | 1·2 V |
| B | 12 V | 12 V | 12 V | 30·7 V | 0 V | 1·2 V |
| C | 12·1 V | 12·1 V | 12·1 V | 11·7 V | 11·4 V | 1·2 V |
| D | 12·2 V | 12·2 V | 12·2 V | 11·7 V | 1·2 V | 1·2 V |
| E | 12 V | 12 V | 12·5 V | 32 V | 1·6 V | 0 V |

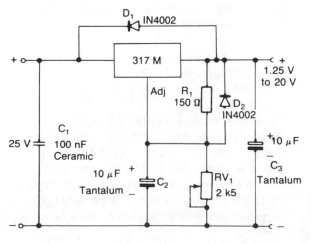

**Fig. 3.18** Variable power supply

## 3.11 Exercise: A variable power supply using a standard IC three terminal regulator (Fig. 3.18)

A wide range of standard IC regulators are now commonly used in electronic system design. These chips have all the necessary series regulator parts, including overload protection, built into one package. Common types include fixed positive and negative regulators such as the 78/79 series and adjustable regulators such as the 723 and the 317M/K. The latter is the one used in this example.

*Circuit specification:*
$V_{in}$: 25 V
$V_{out}$: 1.25 V to 20 V
$I_{0\,max}$: 500 mA
Load regulation: better than 0·1%

The available load current, $I_{0\,max}$, depends upon the power dissipation of the 317M. The current available at various voltages is given by the formula:

$$I_{0\,max} = \frac{Pd}{(V_{in} - V_0)}$$

where $Pd = 7·5$ W and assuming a large heat sink is used for the chip.

Then at $V_{out} = 1.25$ V $I_{0\,max} \simeq 300$ mA

With this IC additional decoupling capacitors are required to ensure that the output ripple and noise are kept to a low value. Protection diodes $D_1$ and $D_2$ prevent damage to the IC in the event of a reverse bias condition.

## Questions

The circuit is loaded with $I_L$ set to 200 mA and the output voltage is adjusted to be about 20 V. Readings were taken with an analogue multimeter. State, with supporting reasons, the faulty component for each of the following conditions.

| Fault | 1 | 2 | 3 |
|---|---|---|---|
| A | 25 V | 1·3 V | 0·1 V |
| B | 25 V | 23·4 V | 23·4 V |
| C | 12 V | 1·2 V | 0 V |
| D | 25 V | 23·5 V | 23·5 V |
| E | 25 V | 23·5 V | 22 V |

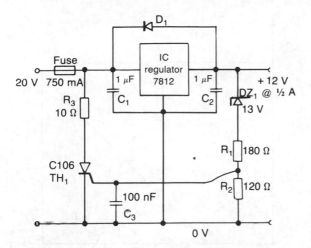

**Fig. 3.19**

## 3.12 Exercise: An overvoltage protection circuit
(Fig. 3.19)

Overvoltage protection becomes an essential feature when a power supply could, in the event of a fault condition, damage load circuits by forcing the supply rail to rise excessively. In this example a 7812 IC regulator is providing a 12 V regulated line to a ½ A load from an input of 20 volts d.c. Suppose the IC fails short circuit. Then the full 20 volts is applied to the load. In this case it is assumed that the load must not be supplied with more than 15 volts, hence the need for the overvoltage protection.

A sensing circuit of $DZ_1$, $R_1$ and $R_2$ detects the overvoltage condition and feeds back a trip signal to the gate of a sensitive thyristor ($TH_1$). This thyristor is triggered into conduction pulling the input voltage to the IC low and then blowing the series fuse.

The trip voltage is given by the formula:

$$V_T \simeq V_{GT} \frac{(R_1 + R_2)}{R_2} + V_{z_1}$$

where $V_{GT}$ is the gate trigger voltage of the thyristor and $V_{z_1}$ is the nominal zener voltage of $DZ_1$.

For the C106 $V_{GT} = 0 \cdot 8$ V
and $I_{GT} = 0 \cdot 2$ mA

Therefore the current through $R_1$ and $R_2$ is made 2 mA and $V_T$ is approximately 15 volts.

### Questions

(1) Explain the purpose of $C_3$.
(2) If $DZ_1$ is a BZY88 13 V regulating diode with a tolerance of $\pm 5\%$ state the upper and lower values of the trip voltage.
(3) Under normal operating conditions what would be the expected voltage at the junction of $R_1$ and $R_2$?
(4) Explain the fault symptoms for
    ($a$)  $R_2$ open circuit
    ($b$)  $DZ_1$ open circuit
    ($c$)  Thyristor anode to cathode short

## 3.13 Exercise: A linear regulator using the 723 IC with foldback current limiting

Most of the earlier examples of power supplies in this chapter have a current limit that operates to restrict the short circuit output current to a value just greater than the full load current. Although simple current limits are effective in providing protection they can lead to a large power loss in the regulator under short circuit output conditions. Consequently, a large heat sink has to be used to prevent the excess heat from destroying the series element.

Suppose we have a regulator with the following specification:

$$\begin{aligned} V_{in} &= 18 \text{ V nominal} \\ V_{out} &= 12 \text{ V} \\ I_L &= 1 \text{ A full load} \\ I_{SC} &= 1 \cdot 2 \text{ A} \end{aligned}$$

Under short circuit output conditions, as shown in Fig. 3.20, the power loss in the regulator is given by:

$$\begin{aligned} P_{tot} &= V_{in} I_{SC} \\ &= 18 \times 1 \cdot 2 \text{ W} \\ &= 21 \cdot 6 \text{ W} \end{aligned}$$

This may be an excessive amount since the normal full power output only causes a series power loss of about 6 W. Suppose, however, that beyond a 'knee' value of current the power supply is forced to output less current so that when a short circuit is applied to the power supply terminals the current supplied is only $0 \cdot 4$ A. The power loss in the regulator is then:

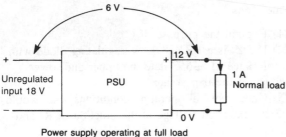

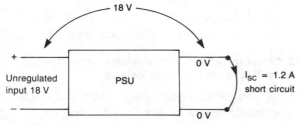

Fig. 3.20   A regulator with simple current limiting

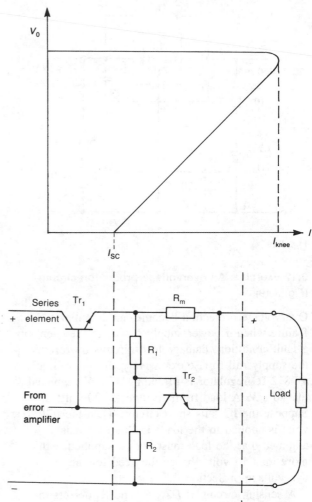

Fig. 3.21   Foldback current limit circuit and characteristics

$$P_{tot} = V_{in} I_{SC}$$
$$= 18 \times 0 \cdot 4 \, \text{W}$$
$$= 7 \cdot 2 \, \text{W}$$

The power loss is now only one-third of the previous value, and the heat sink required for the regulator's series element is much smaller. Another advantage of this type of limiting is that the short circuit current supplied to the load is much lower than the full load current and is therefore unlikely to damage the load. This type of current limit is called 'foldback protection' and has the characteristics and basic circuit outline as given in Fig. 3.21.

The operation of the circuit is as follows. When the load current is increased beyond $I_{knee}$ the voltage across $R_1$, which opposes the voltage developed across $R_m$, falls. In this way $Tr_2$ turns on more as the output voltage falls and diverts more drive current away from the base of the series element. The overall output current, after the value of $I_{knee}$ is exceeded, therefore falls as the output voltage falls and the short circuit output current is set at a low value compared with the maximum output current.

In this exercise, using a 723 regulator IC, the values of $I_{0max}$ and $I_{SC}$ have been deliberately set at small levels because the 723 IC has only a small power dissipation rating, typically 660 mW at 25 °C, and this value must not be exceeded by the product of $V_{in}$ and $I_{SC}$. However, the circuit clearly illustrates the principle of foldback current limiting. If higher output currents are needed from the circuit it can easily be modified by including an external pass transistor as shown in Fig. 3.22(d).

The IC (723) has inside all the parts necessary to create a working regulator. These are:

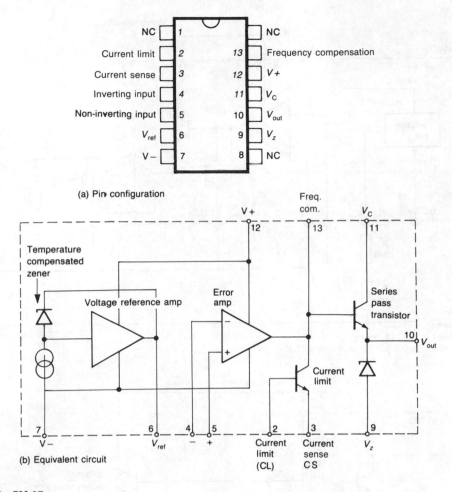

(a) Pin configuration

(b) Equivalent circuit

**Fig. 3.22**  Using the 723 IC

* a $7 \cdot 15$ V temperature compensated reference
* an error amplifier
* a series element
* a limiting transistor

All the user has to do is to fit external resistors and one capacitor. The capacitor (500 pF) prevents unwanted oscillations.

In this design the specification for the power supply is:

$$V_{in} = 10 \text{ V nominal}$$
$$V_{out} = 6 \text{ V}$$
$$I_{0max} = 110 \text{ mA}$$
$$I_{knee} = 100 \text{ mA}$$
$$I_{SC} = 60 \text{ mA}$$

$R_1$ and $R_2$ form a potential divider from the internal reference to create approximately 6 V which is applied to the non-inverting input of the error amplifier. The other input of this amplifier is connected to the power output pin. $R_3$, $R_4$ and $R_5$ form the foldback current limit and set the values of $I_{knee}$ and $I_{SC}$ to those specified.

The test voltages, measured in this case with a DMM and at full load current, are:

| TP | 1 | 2 | 3 | 4 | 5 |
|---|---|---|---|---|---|
| V | $7 \cdot 18$ | $5 \cdot 92$ | $6 \cdot 47$ | $7 \cdot 01$ | $5 \cdot 91$ |

### Questions

(1) Write down the symptoms and test voltage readings for the following faults. The power unit

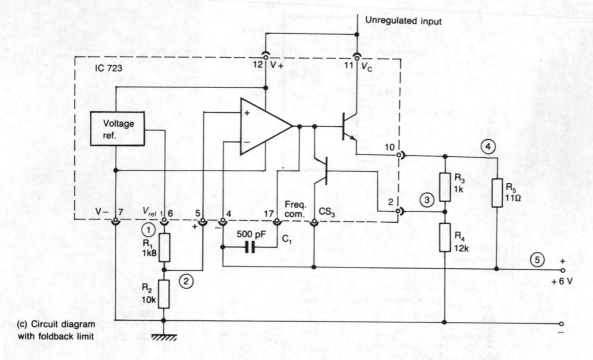

(c) Circuit diagram with foldback limit

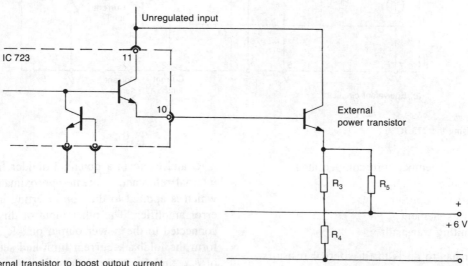

(d) Adding an external transistor to boost output current

**Fig. 3.22** continued

is connected to the normal full load and a DMM on volts range is used.

(a) The IC internal reference fails open circuit
(b) $R_2$ goes open circuit
(c) $R_5$ goes open circuit
(d) The track to pin 12 of the IC goes open circuit

(e) The IC develops a short circuit between pins 2 and 3

(2) State the likely fault symptoms and the effect on the current limit if:

(a) $R_4$ became open circuit
(b) $R_3$ goes open circuit

# 4 Amplifier Circuits

## 4.1 Types and Classes of Amplifier

Since there are so many different types of amplifier used in electronics, before going into circuit details it is a good idea to define what is meant by the term amplifier. An amplifier is any device where a small input signal is used to control a larger output power. It follows from this that an amplifier must consist of some active device, such as a valve or transistor; a source of d.c. power; and a load resistor. This is shown in Fig. 4.1 A. Here the input signal is used to control the current that flows through the active device. This current then develops a voltage change across the load resistor, so that the output power is

$$P_o = V_o i_o \text{ watts}$$

The input power $P_i = V_i i_i$ watts.

POWER GAIN, or power amplification, is given by the ratio of output to input power:

$$A_p = \frac{P_o}{P_i}$$

A more common symbol for the amplifier is shown in Fig. 4.1 B, the signal flow being in the direction of the arrow.

Any amplifier increases the power content of its input signal but this is not always the chief consideration. An amplifier may be designed to give primarily voltage gain, current gain or power gain. Thus the first classification of amplifiers is one which divides them into types designed primarily for power, voltage or current amplification. We have already seen that power gain $A_p = P_o/P_i$. So it follows that

$$\text{VOLTAGE gain } A_v = \frac{V_o}{V_i}$$

$$\text{CURRENT gain } A_i = \frac{i_o}{i_i}$$

It is important to realize that these are all expressions of gain as ratios. In other words if the output

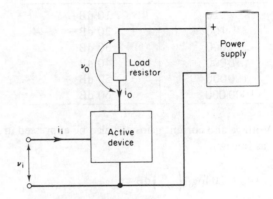

Fig. 4.1A Basic amplifier circuit

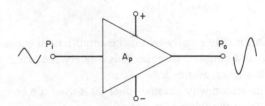

Fig. 4.1B General amplifier symbol
Power gain = $P_o/P_i$

of a voltage amplifier is 2 V peak when its input is 100 mV peak, then voltage gain is

$$\frac{2 \text{ V}}{100 \text{ mV}} = \frac{2 \text{ V}}{0 \cdot 1 \text{ V}} = 20$$

Often the figures involved in working with amplifier gains as ratios can become unwieldy. This is the case when an amplifier exhibits large changes of gain with signal frequency and when these changes have to be recorded graphically. For this reason a logarithmic unit for gain is often used. This is called the Bel.

$$\text{Power gain } A_p = \log_{10}\left(\frac{P_o}{P_i}\right) \text{ Bels}$$

A Bel is usually too large a unit for electronic measurements, so tenths of a Bel or decibels (dB) are commonly used. Then

$$A_p = 10 \log_{10}\left(\frac{P_o}{P_i}\right) dB$$

By using decibels for gain units, very large changes in gain ratios can be compressed. The figures are much easier to handle. This can be seen from the following table.

| Power gain as ratio | Power gain in dB |
|---|---|
| 10 | 10 dB |
| 100 | 20 dB |
| 1 000 | 30 dB |
| 10 000 | 40 dB |
| 100 000 | 50 dB |
| 1 000 000 | 60 dB |

Voltage and current gains can also be expressed in dB, as follows:

$$A_v = 20 \log_{10}\left(\frac{V_o}{V_i}\right) dB$$

$$A_i = 20 \log_{10}\left(\frac{i_o}{i_i}\right) dB$$

This is only strictly true if the amplifier has equal input and output resistance. This is rarely the case, but it is often assumed.

The reason why the multiplier 20 is used can be seen from the following:

$$\text{Power gain} = \frac{P_o}{P_i} = \frac{(V_o/R_o)^2}{(V_i/R_i)^2}$$

where $R_o$ = output resistance and $R_i$ = input resistance.
If $R_o = R_i$, then $A_p = (V_o/V_i)^2$. Therefore

$$\text{Voltage gain} = 10 \log\left(\frac{V_o}{V_i}\right)^2 dB$$

Now $\log x^n = n \log x$

$$\text{Voltage gain} = 2 \times 10 \log\left(\frac{V_o}{V_i}\right) dB$$

As well as possessing gain, any amplifier also has PHASE SHIFT between its output and input. For example at low frequencies a single stage common emitter transistor amplifier produces an output signal which is an inversion of its input. This is shown in Fig. 4.2, and is explained quite simply by

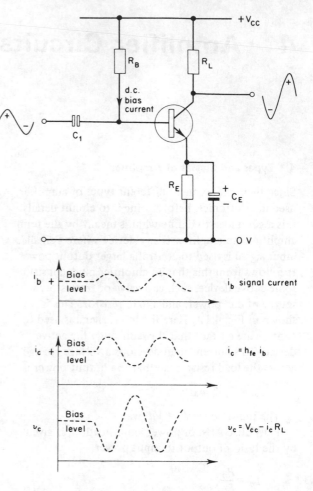

Fig. 4.2 Diagram to explain how inversion takes place between output and input of a single stage common emitter amplifier

the fact that, as the input voltage increases, the current flowing into the base increases, which increases the collector current. This current flows through the collector load and therefore the collector voltage falls. At high signal frequencies, the phase shift does not remain exactly at 180°; this is because (a) the current carriers in the transistor take a finite time to reach the collector region and (b) reactive components in the circuit produce additional phase shifts. Phase shifts in an amplifier at high frequencies lead to phase distortion and possibly instability if a negative feedback loop is used. This will be seen later when negative feedback is discussed.

TABLE 4.1 Broad Classification of Amplifier Circuits

| Gain | Frequency response | Class of operation and typical use |
|---|---|---|
| Voltage, Current or Power | Audio and low frequency Radio frequency (tuned) Wideband or video Pulse Direct current | Class A – Small signal voltage and current amplifiers Class B – Power output amplifiers Class C – Transmitters and pulse switches |

A further classification of amplifiers can be made by considering the RANGE of signal frequency over which the amplifier has useful gain. An audio frequency amplifier, for example, should amplify signals over the range from 15 Hz up to perhaps 20 kHz. A graph of amplifier gain against signal frequency is called a *frequency response curve*. A typical frequency response curve for an audio amplifier is shown in Fig. 4.3. Note that gain is usually plotted in dB on the vertical axis and frequency on the horizontal axis. Frequency is plotted logarithmically so that a large range can be accommodated.

The gain of any amplifier will change because of the reactive components in its coupling and decoupling circuits, stray circuit capacitance and inductance, and because of the frequency limitations of the active devices used.

The BANDWIDTH of an amplifier is usually defined as the range of frequencies over which the gain has not fallen by more than 3 dB from its mid-frequency gain. If the response is flat this is equivalent to 50% of maximum gain for power amplifiers (half power points) and 70·7% of maximum gain for current or voltage amplifiers. From Fig. 4.3 it can be seen that the bandwidth = $f_2 - f_1$.

Amplifiers can therefore be classified as
(a) Audio frequency (AF or LF)
(b) Radio frequency (RF). Tuned with narrow bandwidth
(c) Wideband or video.
(d) D.C. amplifiers.

The basic response curves for these types are shown in Fig. 4.4. For a d.c. amplifier, the active devices must be directly coupled so special techniques are required to ensure correct biasing. This is discussed later.

There is yet a further classification to deal with, namely the CLASS OF OPERATION and the intended use of the amplifier.

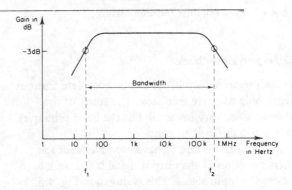

**Fig. 4.3** Typical frequency response curve

There are basically three classes of operation.

Class A   The active device (the transistor or valve) is biased so that a mean current flows all the time. This current is either increased or decreased about this mean value by the input signal. This is the most commonly used class, typical examples being small signal amplifiers.

Class B   The active device is biased just to cut off and is switched into conduction by one half-cycle of the input signal. This class of operation is widely used in push-pull power output amplifiers.

Class C   The active device is biased beyond the point of cut-off so the input signal must exceed a relatively high value before the device can be made to conduct. This class is used in pulse switching and transmitter circuits.

The preceding discussion of amplifier classification can be rather confusing, but it is important before attempting any repair work on an amplifier that the type of amplifier and its purpose is fully appreciated. The information in the last few paragraphs is gathered together in Table 4.1 that sets out the main types of amplifiers in use.

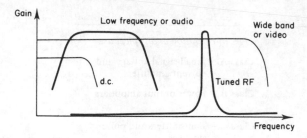

**Fig. 4.4** Various amplifier response curves

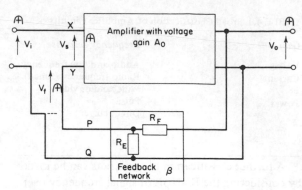

**Fig. 4.5** Block diagram of amplifier with negative feedback

## 4.2 Negative Feedback

No discussion on amplifiers can be complete without mentioning negative feedback. The study of this subject can be very involved, but the basic principles are not too difficult to grasp.

An amplifier is said to have negative feedback when a portion of the output signal is fed back to oppose the input signal. This is shown in Fig. 4.5. Here an amplifier with a gain $A_o$ has a portion of its output signal $V_o$, fed back in series with the input in such a way that it opposes the input. The feedback circuit has a fractional gain of $\beta$. Therefore the feedback signal is given by

$$V_f = \beta V_o \qquad (1)$$

The input signal to the circuit is

$$V_i = V_s + V_f \qquad (2)$$

But substituting (1) in (2) we get

$$V_i = V_s + \beta V_o$$

Now the gain of the amplifier is $A_o = V_o/V_s$. Thus

$$V_o = A_o V_s$$

The gain of the whole circuit, which we shall call $A_c$, is given by

$$A_c = \frac{V_o}{V_i} = \frac{A_o V_s}{V_s + \beta A_o V_s}$$

i.e. $\quad A_c = \dfrac{A_o}{1 + \beta A_o}$

The gain of the amplifier without feedback is called the OPEN LOOP GAIN $A_o$.

The gain of the amplifier circuit with negative feedback is called the CLOSED LOOP GAIN $A_c$.

The *loop* referred to is the connection between the output to the input via the feedback network $\beta$.

The product $A_o\beta$ is called the LOOP GAIN. It is the gain of the circuit from the amplifier input terminals X–Y to the feedback terminals P–Q. Now if the loop gain is much greater than unity, the gain of the amplifier with negative feedback can be re-written as follows:

$$A_c \simeq \frac{A_o}{A_o\beta}$$

(The 1 in the denominator can be ignored since it is small compared with $A_o\beta$) Therefore

$$A_c \simeq \frac{1}{\beta}$$

This is an important result since it means that the gain is now dependent only on the characteristics of the feedback circuit. Thus if the feedback network is a potential divider made by two resistors as shown in Fig. 4.5 then the amplifier gain is

$$A_c \simeq \frac{1}{\beta} \simeq \frac{R_F + R_E}{R_E}$$

The gain is therefore given by a ratio of two resistors, and is independent of changes in circuit components, such as current gain changes in transistors. This holds true as long as the loop gain $A_o\beta \gg 1$.

Negative feedback is widely used for the following reasons:

(a) It stabilizes the gain of the circuit, making the gain independent of changes due to components, temperature and power supply lines.

(b) The frequency response is improved, and the bandwidth widened. This can be seen from Fig. 4.6.

(c) The way in which the feedback signal is derived from the output and applied to the input can be used to modify the input and output impedance of the circuit.

(d) Non-linear distortion and internally generated noise in the amplifier is reduced.

These reasons show that a manufacturer making, say, wideband linear amplifiers, can by using negative feedback ensure that every amplifier produced has nearly the same characteristics.

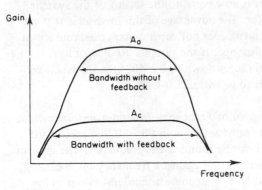

**Fig. 4.6** The effect of negative feedback on the bandwidth of an amplifier

A typical example of a two-stage amplifier using negative feedback is shown in Fig. 4.7. A feedback resistor $R_F$ is connected from the output to the emitter of the first stage. The overall gain of the circuit will be given by

$$A_c = \frac{R_F + R_E}{R_E} \quad \text{i.e. a gain of approximately 30}$$

This type of feedback is said to be *shunt-derived,* since the feedback network is in parallel with the output load, and *series-applied* since the feedback voltage $V_f$ is effectively in series with the input. The output impedance is reduced, while the input impedance is increased.

One of the problems associated with negative feedback is concerned with phase shifts round the loop. In a two-stage voltage amplifier the output is in phase with the input; this is why in Fig. 4.7 the feedback is applied to the emitter and not the base. At higher frequencies the reactive components within

the amplifier introduce additional phase shifts, this in turn changing the phase of the feedback signal. At some frequency the total phase shifts will be such that the feedback signal is adding to input not opposing it. The result is that the circuit will oscillate. This situation can be avoided by ensuring that the loop gain $(A_o\beta)$ is less than unity when the total phase shift round the loop is such as to produce positive instead of negative feedback. This is why op-amps such as the 709 have to have a frequency compensation circuit to limit the bandwidth. For this reason also, direct coupling is often used since it eliminates phase shifts due to coupling capacitor. Fig. 4.8 shows

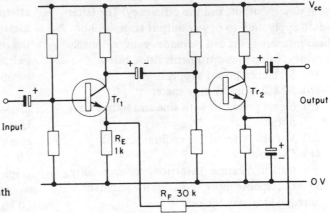

**Fig. 4.7** Two-stage amplifier with negative feedback (a.c. coupled)

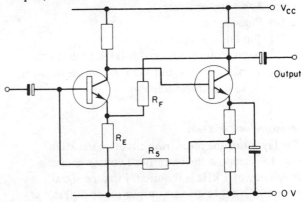

**Fig. 4.8** Two-stage amplifier with negative feedback (d.c. coupled)

the amplifier of Fig. 4.7 using direct coupling. An additional feedback path is provided to stabilize the d.c. operating point via $R_s$.

Feedback can, of course, result from fault conditions when for example a decoupling capacitor goes open circuit. In this case the gain will be reduced drastically.

Testing amplifiers with negative feedback loops is discussed in the next section.

## 4.3 Testing Amplifiers: Basic Measurements

The various tests that should be made on an amplifier system obviously depend upon the type of circuit under consideration. The basic measurements that should be made are those of gain, frequency response, and bandwidth. In addition it may be necessary to measure the input and output impedance, the maximum power output, and the efficiency. The latter would apply only to power output stages. All of these measurements can be made, with reasonable accuracy, using the instruments listed below:

(a) Stabilized power supply
(b) 20 kΩ/V multirange meter
(c) Signal generator with sine and square wave output
(d) Variable attenuator, calibrated in dB.
(e) Oscilloscope.

For tests to measure distortion, noise, stability and pulse response more specialized equipment is required, which may include

(a) Distortion meter
(b) Noise measuring set
(c) Spectrum analyser
(d) Phase meter
(e) Function generator.

It is beyond the scope of this book to detail the highly specialized tests, but the following is intended as a guide to basic measurements.

### Measurement of Gain

The layout of the measuring circuit is shown in Fig. 4.9. Suppose the amplifier's voltage gain at a frequency of 1 kHz is required. First the signal generator is set to give an output of say 500 mV at 1 kHz, with the attenuator switched to zero dB. This signal, at the amplifier input (point A), is connected to the Y-input of the oscilloscope and the oscilloscope controls are adjusted so that the trace displayed uses a large portion of the screen and has its peaks just on graticule lines. The oscilloscope leads are then moved to the amplifier output (point B) and, leaving the

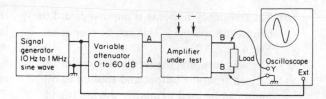

**Fig. 4.9** Laboratory set-up to measure the voltage gain of an amplifier

oscilloscope controls as set, the attenuation is increased until the output is exactly the same height as with the first measurement. The gain of the amplifier is now equal to the setting of the switched attenuator. The advantage of this method is that the measurement does not depend upon the accuracy of the oscilloscope. If the variable attenuator has switched ranges down to 0·1 dB, then the result will be obtained to within ± 0·1 dB.

### Measurement of Frequency Response and Bandwidth

Using the same set-up as in Fig. 4.9, the gain of the amplifier can be found at any frequency. The gain, in dB, is then plotted against a frequency on linear/log graph paper. For an audio amplifier 4 cycles of log would be required to cover the frequency range 10 Hz to 100 kHz.

The bandwidth can be quickly determined by noting the two frequencies at which the gain falls by 3 dB from the mid-frequency gain.

### Measurement of Input Impedance

The input circuit of an amplifier can be represented by a resistor in parallel with a low-value capacitor. At low frequencies the input impedance is mostly resistive since the reactance of the capacitor is such a high value. A circuit for measuring input impedance at 1 kHz is shown in Fig. 4.10. A variable resistor, usually a decade resistance box, is connected between the signal generator and the amplifier input. This

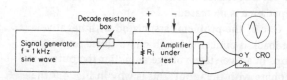

**Fig. 4.10** Measurement of the input impedance of an audio voltage amplifier

resistor is set to zero and the amplifier output is connected to the measuring instrument, an oscilloscope or a.c. meter. The controls are set so that a large deflection is indicated. The resistance of the decade box is then increased until the indicated output signal falls by exactly a half. Since the resistance box and the amplifier input impedance form a potential divider when the output is halved, the setting of the decade resistance box is equal to the input resistance.

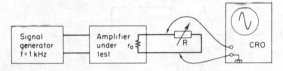

**Fig. 4.11** Measurement of output impedance of a voltage amplifier

### Measurement of Output Resistance

The circuit shown in Fig. 4.11 is used for this measurement. The technique is similar to that of measuring the input impedance. A signal frequency of 1 kHz is used and initially $R_L$ is disconnected and a large deflection obtained on the oscilloscope. The external load $R_L$ is then connected and reduced in value until the output falls by exactly a half. The value of $R_L$ at which this occurs is equal to the resistance.

### Measurement of Power Output, Efficiency, and Sensitivity for an Audio Amplifier

For these measurements the loudspeaker should be replaced by a wire-wound load resistor of the same value as the loudspeaker impedance, and the tests should be carried out at a frequency where the loudspeaker impedance would be mostly resistive, typically about 1 kHz. The diagram for the measurement is shown in Fig. 4.12. The wattage rating of the load resistor should be higher than that of the maximum output power. The input voltage should be adjusted until the output signal indicated by the oscilloscope is a maximum undistorted level. This is when there is no clipping of the positive and negative excursions of the output signal. Naturally if a distortion meter is available then a more accurate check on distortion levels can be made. Then the maximum output power should be recorded without exceeding the manufacturers specified value of

harmonic distortion. This may be a value of total harmonic distortion of 0·05% of the output signal.

$$\text{Power output} = \frac{V_o^2}{R_L}$$

where $V_o$ is the r.m.s. value of the output signal.

$$\text{Remember} \quad \text{r.m.s.} = \frac{\text{peak to peak value}}{2\sqrt{2}}$$

The efficiency of the amplifier can be checked by measuring the d.c. power taken by the amplifier from the supply.

$$\text{D.C. power} = V_{dc}I_{dc}$$

and

$$\text{Power efficiency} = \frac{\text{r.m.s. output power}}{\text{d.c. input power}} \times 100\%$$

The *sensitivity* of the amplifier is the input voltage required at the input to produce maximum undistorted output power.

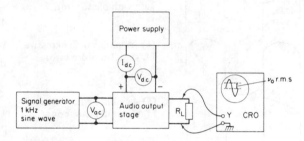

**Fig. 4.12** Measurement of power output, efficiency and sensitivity of an audio output stage.
$R_L$ is a wire-wound resistor of the same value as the loudspeaker impedance.

## 4.4 Transient Testing of Amplifiers

All the tests previously described are made using an input signal at one frequency. By applying pulses or square waves to an amplifier it is possible to acquire information about the amplifiers frequency response, phase distortion, and any tendency to instability.

A square wave is made up of a series of pure sine wave components, which are a fundamental, having the same periodic time as the square wave, and all odd harmonics. Thus by applying a square wave or pulse to an amplifier, a large range of signals at different frequencies have to be amplified by the same ratio and without phase shift if the output is to be a perfect replica of the input.

For testing low frequency amplifiers, a square wave of 40 Hz or 1 kHz is suitable and the output signal can be observed on an oscilloscope. Departure from squareness in the output signal gives a good indication of the transient distortion that is present in the amplifier. Various conditions are shown in Fig. 4.13.

Video and wideband amplifiers can also be usefully tested in this way, but usually a special form of signal called a pulse and bar is used. This, together with various possible outputs, are shown in Fig. 4.14.

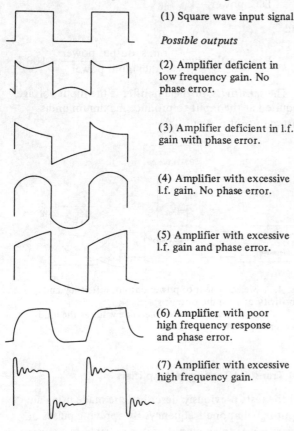

(1) Square wave input signal

*Possible outputs*

(2) Amplifier deficient in low frequency gain. No phase error.

(3) Amplifier deficient in l.f. gain with phase error.

(4) Amplifier with excessive l.f. gain. No phase error.

(5) Amplifier with excessive l.f. gain and phase error.

(6) Amplifier with poor high frequency response and phase error.

(7) Amplifier with excessive high frequency gain.

Fig. 4.13 Square wave testing of an amplifier

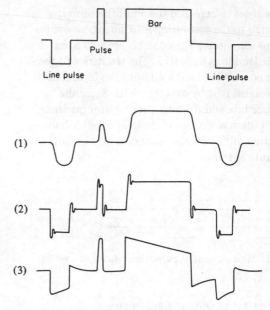

Fig. 4.14 Pulse and bar testing of video amplifiers
(1) Lack of high frequency gain
(2) Excessive high frequency gain
(3) Poor low frequency response

### 4.5 Distortion Measurements

Various types of distortion can effect the shape of the output signal from an amplifier.

#### Amplitude Distortion
The output signal is flattened on one or both of its peaks as shown in Fig. 4.15. This type of distortion occurs when the amplifier is overdriven by an excessively large input signal, or when the bias conditions change, or because of some non-linearity in the characteristics of a transistor or valve.

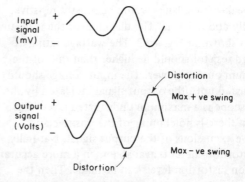

Fig. 4.15 Amplitude distortion resulting from an overdriven amplifier

## Frequency Distortion

This results when the amplifier gain changes drastically with frequency within its passband. Suppose an amplifier has a frequency response as shown in Fig. 4.16, which is reasonably flat over the passband, but that the actual response is as shown in Fig. 4.16B, then the amplifier is said to have frequency distortion. This can take the form of loss of gain at low or high frequencies or increase of gain at low or high frequencies.

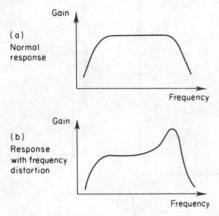

(a)
Normal
response

(b)
Response
with frequency
distortion

**Fig. 4.16** Frequency distortion

## Phase Distortion

As the signal frequency is increased so the phase of the output signal relative to the input will change. This type of distortion is troublesome when the input signal is a complex waveform, made up of several sine wave components all at different frequencies. If these all suffer different phase shifts through the amplifier, the resulting output will not be identical in shape to the input.

## Cross-over Distortion

Distortion of this type occurs in class B push-pull output stages. In a complementary transistor output stage for example, unless some forward bias is applied, the transistors will not conduct until the input signal to their bases exceeds about 500 mV (this is for a silicon transistor). See Fig. 4.17. The purpose of the

**Fig. 4.17** Cross-over distortion

bias components is to overcome the distortion by providing a very small amount of forward bias.

## Intermodulation Distortion

When non-linearity exists in an amplifier circuit, two signals of different frequencies, say 400 Hz and 1 kHz, as well as being amplified will be mixed, and so the output will contain small amplitude signals of the sum and difference frequencies, i.e. at 600 Hz and 1·6 kHz and harmonics of these frequencies.

Measurement of distortion levels is usually made using a distortion meter, an instrument which sums the power in all the harmonics and gives the result as a percentage of the output power. This gives the value of the *total harmonic distortion* resulting from amplitude and non-linear distortion, but does not include frequency, phase or intermodulation distortion. A frequency of 1 kHz is normally used for this measurement.

Total harmonic distortion can also be measured by passing the output voltage signal through a filter which attenuates the measurement frequency (1 kHz) but passes all harmonics. A good circuit for this is a twin-tee filter as shown in Fig. 4.18 since this has maximum attenuation at one frequency. The output can be measured using a sensitive r.m.s. millivolt-meter.

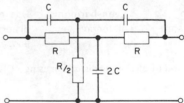

**Fig. 4.18** Twin-tee filter

Intermodulation distortion can be measured by feeding two signals of 400 Hz and 1 kHz into the amplifier usually with a ratio of about 4:1. Then using a filter at 1 kHz the result of any intermodulation will be indicated using the method detailed previously.

A method that can be used to display amplitude distortion, phase shift distortion and harmonic distortion for an audio amplifier is shown in Fig. 4.19. The signal generator set to 1 kHz is fed to the amplifier input at a suitably low level and to the X-input of the oscilloscope. The output from the amplifier is fed to the Y-input of the oscilloscope. The oscilloscope trace will be a straight line at an angle of 45° if the amplifier output is undistorted. Naturally a high-quality oscilloscope must be used for this test, since any non-linearity in the X and Y oscilloscope amplifiers will also be displayed. Various outputs for different types of distortion are shown in Fig. 4.19 also.

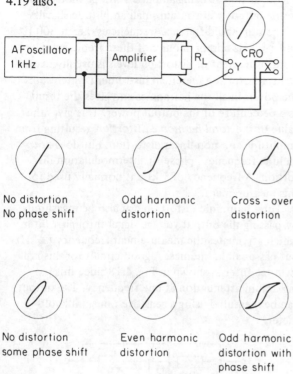

Fig. 4.19 Method of displaying distortion using a CRO

## 4.6 Faults in Amplifiers

It would not be possible to detail all the possible faults that could occur in all the various types of amplifier circuits. Instead the following is a general guide to assist in fault location. Before considering some typical faults it should be noted that, as well as the d.c. bias levels, the output signal itself is often an invaluable guide to the type of fault.

The previous section detailed the types of distortion that could occur. Let's consider a simple example of how changes in bias components can cause large amounts of amplitude distortion. In Fig. 4.20A the operating point at the collector is about +5 V to allow equal positive and negative swings at the output. If $R_1$ goes high in value from

its nominal 82 kΩ to say 150 kΩ the operating point will now rise to approximately 8 V. The output signal is now distorted on its positive excursion as shown in Fig. 4.20B.

For fault finding on amplifier systems it is best to follow the standard procedure and inject a signal into the input, and by using an a.c. meter or an oscilloscope, check each stage in turn until the

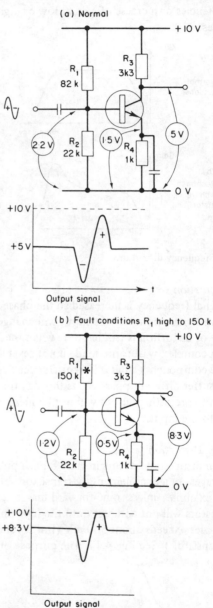

Fig. 4.20 Distortion resulting from a change in bias conditions

faulty stage is reached. Then measure the d.c. levels at this tage.

Table 4.2 lists some typical faults together with the expected symptoms.

## TABLE 4.2 Typical Faults on Amplifier Systems

**Small signal amplifiers**

| FAULT | SYMPTOMS |
|---|---|
| Bias component failure open circuit or high value resistors | Results in a large change in operating point usually tending to cut transistors off. This gives either grossly distorted output or no output at all. |
| Short circuit decoupling or coupling capacitors | Again a large change in operating point usually tending to force transistors to conduct much harder. Grossly distorted output. |
| Coupling capacitors open circuit | No transfer of signal from one stage to next. All d.c. bias levels normal. No output signal. |
| Signal decoupling capacitors open circuit | Low gain, since series negative feedback is introduced. |
| Power line decoupling capacitors open circuit | Increase in "hum" level (100 Hz) at amplifier output. The first stage of a pre-amplifier is normally supplied from a decoupled line. |
| Open circuit feedback line | Excessive gain with instability and possibly oscillation. |
| Noisy transistor or resistor at input | Poor signal-to-noise ratio. (Always check early stages first.) |
| Change in coupling and decoupling capacitor values to lower value | Reduction in bandwidth. Poor low frequency response. |

**Power amplifiers**

| FAULT | SYMPTOMS |
|---|---|
| Bias resistors open circuit or high in value | For class B amplifiers, the type in common use, there will be a large amount of cross-over distortion. |
| Output capacitor short circuit | Output fuses blown or transistors overheating. Use resistance check to find faulty component. |
| Bias potentiometer incorrectly set | Either (i) increase in cross-over distortion or (ii) over-heating of output transistors. |

## 4.7 Exercise: Two-stage Pre-amplifier (Fig. 4.21)

*Specification:*

| | |
|---|---|
| Voltage gain | 34 |
| Input impedance | 100 k$\Omega$ |
| Output impedance | 500 $\Omega$ |
| Frequency response | 20 Hz to 30 kHz |
| Sensitivity for 1 V r.m.s. output | 30 mV r.m.s. |

This circuit, using negative feedback loops to stabilize both the a.c. gain and the d.c. operating point, is a typical example of a design as discussed in section 4.2.

A low noise high gain transistor is used for $Tr_1$, and since series negative feedback is used the input impedance is almost equal to $R_3$, which is 100 k$\Omega$.

The a.c. gain of the circuit is determined by the negative feedback loop from $Tr_2$ collector via $R_5$ to the emitter of $Tr_1$ and $R_2$. Since $R_5$ and $R_2$ form a simple potential divider, the feedback fraction $\beta$ is given by

$$\beta = \frac{R_2}{R_2 + R_5}$$

Now if the loop gain $A_o\beta$ is much greater than unity, as is the case in this circuit, the gain with feedback is given by

$$A_c \simeq \frac{1}{\beta} = \frac{R_2 + R_5}{R_2} = 34$$

When you build this circuit, you can easily verify by measurement that the gain is 34, and the open loop gain can be measured by decoupling $R_2$ with a large-value capacitor. This reduces the feedback signal to zero.

Since the circuit uses direct coupling the calculation of the d.c. bias levels is relatively difficult. In designing such a circuit one starts with the required operating point at $Tr_2$ collector. To give undistorted output this should be about half the supply voltage i.e. 7·5 V If we assume that $R_4$, $Tr_2$ collector load is 2 k$\Omega$, $R_5$ is 33 k$\Omega$ and $R_2$ 1 k$\Omega$ to give the voltage gain required then the other component values can be calculated as follows.

(*a*) Current through $R_4 = \dfrac{V_{CC} - V_{C2}}{R_4} = \dfrac{7\cdot5}{2\,k\Omega} = 3\cdot4$ mA

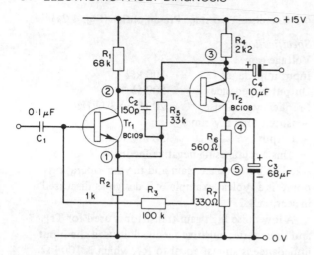

**Fig. 4.21** Two-stage pre-amplifier

(b) Current through $R_5 = \dfrac{V_{C2}}{R_2 + R_5} = \dfrac{7 \cdot 5}{34 \text{ k}\Omega} = 0 \cdot 22$ mA

and the current through $Tr_2$ collector $= 3 \cdot 4 - 0 \cdot 22$

$= 3 \cdot 18$ mA

(c) Assuming that $Tr_1$ current is low, say $100 \,\mu$A by design, then the voltage at $Tr_1$ emitter is given by

$$V_E = R_2 \, (I_E + I_{R_s})$$

$$= 1 \text{ k}\Omega \, (0 \cdot 1 + 0 \cdot 22) \text{mA} = 320 \text{ mV}$$

(d) Voltage at $Tr_1$ base $= V_E + V_{BE} \simeq 1$ V and if we make the simplifying assumption that $Tr_1$ base current can be neglected then the voltage at the junction of $R_7$ and $R_6$ will also be 1 V. The value for $R_7$ is now given since

$$R_7 = \frac{V_{R_7}}{I_{E2}} = \frac{1}{3 \cdot 18} \text{ k}\Omega = 314 \ \Omega$$

$$(330 \text{ n.p.v.})$$

(e) By making $R_6$ a 560 $\Omega$ resistor the voltage at $Tr_2$ emitter will be about $2 \cdot 8$ V.

$$V_{E2} = (R_6 + R_7) \, I_{E2}$$

and the base voltage of $Tr_2$ is

$$V_{E2} + V_{BE} = 2 \cdot 8 + 0 \cdot 7 = +3 \cdot 5 \text{ V} = V_{C1}$$

This will be the voltage at $Tr_1$ collector which should allow a reasonable voltage swing at that point.

(f) We made $I_{C1}$ to be $100 \,\mu$A in the design, but in addition to this current flowing through $R_1$ there is also the base current for $Tr_2$. Now

$$I_{B2} = \frac{I_{C2}}{h_{FE(min)}} = \frac{3 \cdot 18 \text{ mA}}{60} = 53 \,\mu\text{A}$$

Therefore

$$R_1 = \frac{V_{CC} - V_{C_1}}{I_{C_1} + I_{B_2}} = \frac{15 - 3 \cdot 5}{153 \,\mu\text{A}} = \frac{11 \cdot 5}{0 \cdot 153} = 75 \text{ k}\Omega$$

A value of 68 k$\Omega$ was chosen.

The d.c. levels at the various test points are shown for the calculated and measured values below. A 20 k$\Omega$/V meter was used.

| TP | 1 | 2 | 3 | 4 | 5 |
|---|---|---|---|---|---|
| Calc-ulated | 0·32 | 3·5 | 7·5 | 2·8 | 1 |
| Meas-ured | 0·39 | 3·4 | 7·5 | 2·8 | 1·06 |

It is important to realize that the d.c. operating point is stabilized by both feedback loops and that any tendency for the operating point to change would be counteracted by more or less d.c. feedback.

$C_2$ is included to limit the high frequency gain; above 30 kHz its reactance is less than $R_5$ and therefore the feedback signal increases, thereby reducing the overall gain.

Let's consider the effect of some component failures on the operation of this circuit. Imagine that the d.c. feedback loop via $R_3$ goes open circuit. With $R_3$ open there can be no bias current into $Tr_1$ and thus $Tr_1$ is cut off. Normally we would expect the collector voltage of a transistor that is cut off to rise to the supply voltage, but in directly coupled circuits this is not possible. In this case, $R_1$ feeds bias current to $Tr_2$, so with $Tr_1$ off the bias to $Tr_2$ increases and we must expect $Tr_2$ collector voltage to fall and its emitter voltage to rise.

The actual readings with $R_3$ open circuit are:

| TP | 1 | 2 | 3 | 4 | 5 |
|---|---|---|---|---|---|
| Meter reading | 0·15 | 5 | 4·3 | 4·25 | 1·6 |

It is interesting to note that these readings also result if $Tr_2$ develops either a short circuit base emitter junction, or if $Tr_2$ collector is open circuit. Verify these for yourself.

Consider now the effect of a base emitter short on $Tr_2$. Obviously test points (2) and (4) will give the same reading, and we would expect test point (3) to rise since $Tr_2$ can no longer act as a transistor and therefore collector current ceases. Since $R_1$ is such a high value in comparison to $R_6$ and $R_7$, the voltage at (2) and (4) will be low. It can be quickly calculated since $R_1$, $R_6$ and $R_7$ will form a potential divider.

Test point (4) and (2) with $Tr_2$ b.e. shorted is

$$V_4 = \frac{15 \times (R_6 + R_7)}{R_1 + R_6 + R_7} = \frac{15 \times 0.89 \text{ k}\Omega}{68.89 \text{ k}\Omega} = 194 \text{ mV}$$

Test point (5) will be then about 70 mV.

Since $Tr_2$ is passing no collector current, test points (3) and (1) will also be determined by the potential divider formed by $R_4$, $R_5$ and $R_2$. Try and calculate the values you would expect to measure before looking at the next table.

In fact the measured values with $Tr_2$ base emitter short are:

| TP | 1 | 2 | 3 | 4 | 5 |
|---|---|---|---|---|---|
| MR | 0.4 | 0.2 | 14 | 0.2 | 0.08 |

**Questions** All the readings were made using a standard 20 k$\Omega$/V multirange meter. State which component or components could cause the symptoms and give details of the type of component fault. The output was observed on an oscilloscope with an input of 30 mV r.m.s. at 1 kHz.

| Fault | 1 | 2 | 3 | 4 | 5 | Output signal |
|---|---|---|---|---|---|---|
| A | 0 | 0.75 | 0.1 | 0 | 0 | Zero output |
| B | 0.2 | 2.5 | 9.2 | 2.3 | 0.9 | Output 12 V pk-pk gross distortion |
| C | 0.6 | 0.6 | 14 | 0.06 | 0.02 | Zero output signal |
| D | 2.7 | 3.3 | 2.7 | 2.65 | 1 | Zero output signal |
| E | 0.15 | 0.8 | 14 | 0.2 | 0.08 | Zero output signal |
| F | 0.39 | 3.4 | 7.5 | 2.8 | 1.06 | Much reduced output signal amplitude |
| G | 0.42 | 0 | 14 | 0 | 0 | Zero output signal |

### 4.8 Exercise: Pre-amplifier with a FET Input (Fig. 4.22)

One of the important features of Field Effect Transistors is that under normal operating conditions they have a very high input impedance. This makes them ideal devices for amplifying signals from transducers such as the piezo-electric crystal and the ceramic pick-up, both of which must work into a high resistance. The crystal pick-up can be simply considered as a low-value capacitor with a small voltage generator in series (Fig. 4.23). The voltage is induced across the crystal when a mechanical force is applied to it from the modulated groove on the record. At the low-frequency end of the audio spectrum the small series capacitor has a high reactance (remember $X_c = 1/2\pi f_c$) so that if the input resistance of the pre-amplifier is not high, the low frequencies will be attenuated and the bass response lost.

Another useful advantage of using a FET at the input stage is that the noise generated is less than that from a bipolar transistor. FETs use only one type of charge carrier and this tends to result in lower electrical noise. In amplifier systems a low-noise device is usually required at the input stage since this is where noise generated within the amplifier will receiver the greatest amplification. The noise factor of a device is a direct measure of how much noise is added to the input signal. Suppose, for example, that the input signal to an amplifier has a signal noise ratio of 40 dB (a voltage ratio of 100:1), then if the noise factor of the amplifier is 4 dB, the resulting signal-to-noise ratio of the amplifier's output will be 36 dB. Noise factors for FETs can be as low as 2 dB.

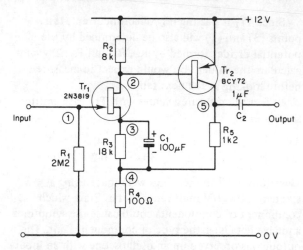

**Fig. 4.22** FET input pre-amplifier

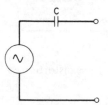

**Fig. 4.23** Equivalent circuit of crystal pick-up

The *specification* for the pre-amplifier is

| | |
|---|---|
| Voltage gain | 12 |
| Input impedance | 2 MΩ at audio frequencies |
| Output impedance | 1300 Ω |
| Frequency response | 20 Hz to 30 kHz |
| Output signal | 1 V r.m.s. |

A low voltage gain is used since the signal from the pick-up is usually 100 mV or so.

An n-channel junction FET (2N3819) is used as a common source amplifier in the input stage, and this is directly coupled to $Tr_2$, a common emitter amplifier which provides most of the gain.

The FET is provided with self-bias via $R_3$ which works in the following way. The gate is connected to the 0 V rail via $R_1$ so that when power is applied the gate-to-source voltage is initially zero; this causes a drain current to flow, which in turn causes a voltage to be developed across $R_3$. The source voltage rises positive, thus increasing the gate-to-source reverse bias and causing the drain current to remain constant at a reasonable value.

Negative feedback is provided by $R_5$ and $R_4$. Actually the feedback signal is voltage-derived, since $R_4$ is in series with the output resistor $R_5$, and series-applied since $R_4$ is also in series with $R_3$. The gain is given approximately by the ratio of $R_5$ and $R_4$.

$C_1$ is a decoupling capacitor which is necessary to ensure that a high overall open loop gain is achieved. Its value determines the low-frequency gain point.

When you build this circuit you may find that the d.c. bias voltages are not the same as those given in the next table. This is because the parameters of FETs do differ widely. What is important is that the operating point of $Tr_2$ collector (point 5) should be such as to allow reasonable positive and negative swing at the output. This output signal is suitable for driving a medium power output stage; a volume control would be required.

The voltages at the d.c. test points measured with a 20 kΩ/V meter with respect to 0 V were

| TP | 1 | 2 | 3 | 4 | 5 |
|---|---|---|---|---|---|
| MR | 0 | 11·3 | 3 | 0·6 | 8 |

**Questions**    The next set of readings were taken under fault conditions. Identify, with reasons, the faulty component and the type of fault.

| | 1 | 2 | 3 | 4 | 5 | Output |
|---|---|---|---|---|---|---|
| Fault | | | | | | |
| A | 0 | 11·9 | 3·3 | 3·2 | 3·2 | Zero |
| B | 0 | 11·2 | 1·2 | 1·2 | 11·9 | Zero |
| C | 0 | 11·3 | 3 | 0·6 | 8 | Very low gain |
| D | 0 | 12 | 0 | 0 | 0 | Zero |
| E | 0 | 10·5 | 3 | 0 | 0·1 | Zero |

### 4.9 Exercise: D.C. Amplifier (Fig. 4.24)

D.C. amplifiers are used to amplify slowly varying signals from transducers such as thermocouples, thermistors, strain gauges, photocells, etc. As well as amplifying the signal, the amplifiers' outputs must not drift. Drift in d.c. amplifiers is defined as any change in output signal when the input is short circuited or held at zero. Drift is caused by several factors but the most important are those that effect the input stage; these are temperature and power supply voltage changes. The latter can be minimized by using a well stabilized power unit so we shall consider the effect of temperature. This affects transistors in three main ways: firstly it changes the parameters, namely the current gain, secondly it alters the leakage current, and thirdly it changes the base emitter forward bias voltage.

Changes in $h_{FE}$ with temperature can be minimized by using negative feedback. Leakage currents can be kept low by using silicon planar transistors, and so we are left with the main cause of drift being the change of $V_{BE}$ with temperature. In any transistor, $V_{BE}$ changes by approximately $-2$ mV per degree C rise in temperature. This may not sound a lot, but imagine that a single transistor is used as the input stage of a d.c. amplifier with a gain of 25. Then for every one degree change in temperature, the output will drift by 50 mV because the 2 mV change of $V_{BE}$ at the input transistor will appear as a d.c. input signal. For this reason single transistors are rarely used as the input stage of d.c. amplifiers. The circuit in common use is the differential amplifier shown in Fig. 4.25, in which two transistors are wired together in a balanced arrangement. For this type of circuit two sets of input conditions can be considered:

  (i)  Input of opposite polarity – called differential mode inputs.

  (ii)  Inputs of the same polarity – called common mode inputs.

For the differential mode input signals a large output signal is generated. Imagine point A going positive while B goes negative, $Tr_1$ will conduct more than $Tr_2$ and the output signal will be large.

A common mode signal on the other hand will result in little change at the output since both transistors conduct more or less equally. Now as long as the two transistors are matched and situated close together, preferably in the same enclosure, then any changes in the $V_{BE}$ of both transistors with temperature can be considered as a common mode input, with the result that little change in the output signal takes place. Thus by using the differential amplifier, drift with temperature can be kept to a low value.

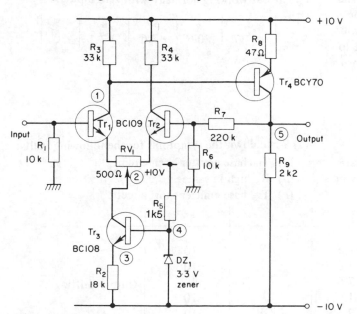

**Fig. 4.24** D.C. amplifier

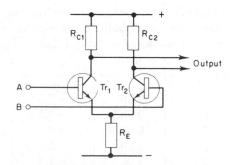

**Fig. 4.25** Basic differential amplifier

A measure of the quality of a differential amplifier is called common mode rejection ratio (CMRR)

$$CMRR = \frac{Differential\ gain}{Common\ mode\ gain}$$

To achieve high values of CMRR it is necessary to make the emitter resistor $R_E$ as high in value as possible, since it is this resistor that provides the negative feedback which keeps the common mode gain low. For this reason the current to the differential amplifier is often supplied from a constant current source, and $R_E$ is then equal to the high slope resistance of the output of this source. In the example (Fig. 4.24), $Tr_3$, wired as a common emitter amplifier, supplies the constant current.

The d.c. amplifier in this exercise is relatively simple and, therefore, does not produce the best results in terms of drift and stability that can be achieved. Nevertheless, if care is taken over selection of the transistors and in construction of the circuit, fair performance is possible.

The specification is

| | |
|---|---|
| Voltage gain | 22 |
| Input resistance | 10 kΩ |
| Output voltage | ± 5 V maximum |
| Temperature drift | ± 3 mV per °C |
| Stability | ± 10 mV per hour |

$Tr_1$ and $Tr_2$ form the differential amplifier, and these transistors should be carefully matched for current gain and mounted close together. This will ensure the best drift performance. It is also wise to shield the circuit from draughts by building it into a small enclosure. When you test the circuit you can observe the effects of temperature by blowing on one of the transistors and noting the large change in d.c. output. A constant current of about 150 $\mu$A is provided from $Tr_3$. This current is determined by the zener diode and $R_2$.

The output from $Tr_1$ collector is connected to the common emitter amplifier $Tr_4$. The gain of the circuit is stabilized by a negative feedback loop from the collector of $Tr_4$ via $R_7$ to the base of $Tr_2$. The potentiometer $RV_1$ in the differential amp, is used to offset any differences between $Tr_1$ and $Tr_2$ base potentials when the input is zero. Under these conditions it should be adjusted to give zero output volts.

The drift and stability of the amplifier is best checked by a chart recorder. The voltage gain can be measured by applying d.c. signals from a stable millivolt source.

In operation a small positive input signal (d.c.) will cause $Tr_1$ to conduct more heavily than $Tr_2$, thus the collector voltage of $Tr_1$ will fall causing $Tr_4$ to conduct more. The output voltage rises, and a portion of this is fed back to $Tr_2$ base which opposes the input. To alter the gain, change the ratio of $R_7$ and $R_6$.

The voltages at the test points with respect to 0 V were measured using a digital meter when the circuit was working normally with zero input.

| TP | 1 | 2 | 3 | 4 | 5 |
|---|---|---|---|---|---|
| V | 9.17 | −0.6 | −7.6 | −7.1 | 0 |

## Questions

(1) From the following set of readings, which were taken under fault conditions, determine which component is faulty and its type of fault.

(2) Write down the symptoms for the following faults:
  (a) $Tr_2$ base emitter open circuit.
  (b) $R_5$ high in value.
  (c) $Tr_3$ base emitter short circuit.

| Fault | 1 | 2 | 3 | 4 | 5 | |
|---|---|---|---|---|---|---|
| A | 9.98 | +7.3 | +7.2 | +7.9 | −10 | |
| B | 9.98 | −7.5 | −7.6 | −7.1 | −10 | |
| C | 9.05 | −0.6 | −7.6 | −7.1 | +0.2 | Output drifting |
| D | 9.98 | 0 | −10 | −10 | −10 | |
| E | 9.98 | −1.1 | −7.6 | −7.1 | −10 | |

## 4.10 Exercise: Audio Power Amplifier (Fig. 4.26)

*Specification*

| | |
|---|---|
| Power output | 4 W into 8 Ω |
| Harmonic distortion | Less than 2% at max output |
| Sensitivity | approximately 1 V r.m.s. |
| Frequency response | 15 Hz to 20 kHz |
| Input impedance | 1k5 Ω |

The circuit uses a standard class B complementary output stage with a matched pair of medium power transistors (type BD131 and BD132). These two transistors are fed via transistors $Tr_2$ and $Tr_3$ from the output of a common emitter amplifier $Tr_1$. $Tr_2$, $Tr_4$ and $Tr_3$, $Tr_5$ are wired as complementary pairs with very high current gains.

Transistors $Tr_2$ and $Tr_4$ conduct on the positive half-cycles of the signals across $R_4$, while $Tr_3$ and $Tr_5$ conduct on the negative half-cycles. Naturally a small amount of forward bias must be supplied to the output transistors, otherwise the output signal would contain cross-over distortion. This bias is provided by diodes $D_1$ and $D_2$, and the low-value variable resistor $RV_1$. To enable a small quiescent current to flow through $Tr_4$ and $Tr_5$ a d.c. voltage of approximately 1·25 V must be set up between the base connections of $Tr_2$ and $Tr_3$. The diodes $D_1$ and $D_2$, being forward biased by $Tr_1$ collector current, give about 1·4 V and $RV_1$ is used to adjust the level. This potentiometer must be adjusted carefully during the initial setting up so that the current flowing in $Tr_5$ emitter is about 10 mA. To do this the link A is broken and a d.c. milliammeter placed in circuit to monitor the current. This setting should remove any tendency for cross-over distortion to occur.

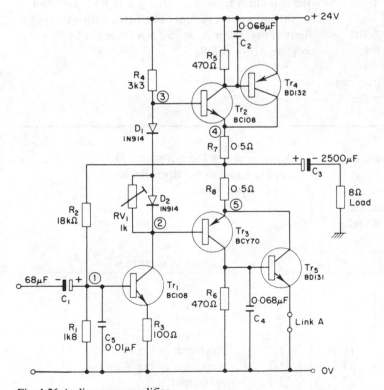

Fig. 4.26 Audio power amplifier

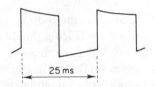

Fig. 4.27 Square wave test on audio amplifier Input 40 Hz square wave

Fig. 4.28 Wave form for fault D ($f = 1$ kHz)

Fig. 4.29 Wave form for fault E ($f = 1$ kHz)

The two output transistors must be mounted on a heat sink, about 100 cm$^2$ of 3 mm thick aluminium is sufficient. When the output transistors warm up during high power output, or when the ambient temperature changes, changes in the base emitter voltages of $Tr_2$ and $Tr_3$ take place. The diodes $D_1$ and $D_2$ change with temperature also so that the overall bias conditions do not alter. $D_1$ and $D_2$ then provide a degree of thermal compensation.

A stabilizing negative feedback loop is provided via $R_2$, which feeds a portion of the output signal back to oppose the input. Since the circuit is directly coupled, the feedback also stabilizes the d.c. operating point at the junction of $R_7$ and $R_8$. If the circuit is working correctly, $Tr_2$ matched with $Tr_3$, and $Tr_4$ matched with $Tr_5$, then this d.c. operating point should be exactly half the supply voltage. The rest of the d.c. voltages can be calculated by working from this value and remembering that the voltage between base and emitter of a working transistor should be approximately 0·7 V. The actual voltages with a supply rail of +24 V are

| TP | 1 | 2 | 3 | 4 | 5 | Supply |
|----|-----|------|------|------|------|--------|
| MR | +1 | 11·1 | 12·5 | 11·8 | 11·8 | (24) |

Capacitor $C_3$, the coupling capacitor from the output to the 8 Ω load, should have a working voltage of 25 V and a ripple current rating of at least 600 mA. (Do not fit too small a capacitor, and make sure that it is mounted well away from the heat sinks.)

Since transistors with high values of cut-off frequency are used the circuit may well tend to burst into oscillation. Components $C_2$, $C_4$ and $C_5$ are included to prevent this.

When measuring the power output, an oscilloscope is used to check that an undistorted voltage waveform is set up across the load, and a standard multimeter on a.c. voltage range to measure the r.m.s. value of the output voltage. Then

$$\text{Power output} = \frac{V_o^2}{R_L}$$

where $R_L$ is the 8 Ω load resistor.

The frequency response results are shown in Fig. 4.27 for a 40 Hz square wave input. Note that the low frequency components present in the 40 Hz square wave have less amplification than the higher frequencies. This is not necessarily bad since the lower cut-off frequency of the amplifier needs to be higher than the bass resonance of the loudspeaker.

In Class B output stages, the output transistors are prone to damage if either the output is accidentally short circuited or if an open circuit occurs in the bias chain formed by $D_1$, $D_2$ and $RV_1$. To avoid this the power supply could be fitted with a current limit of say 750 mA, or 750 mA fuses can be fitted in place of $R_7$ and $R_8$.

## Questions

(1) All the following d.c. voltage readings were taken when a fault existed, with an input of 100 mV at 1 kHz. Test points were measured with respect to 0 V using a standard multimeter. Consider each in turn and try and deduce which component is faulty and its type of fault.

(2) What would be the symptoms for
 (a) $Tr_4$ base emitter open circuit
 (b) $C_3$ open circuit.

| | 1 | 2 | 3 | 4 | 5 | Additional symptom |
|---|------|------|------|------|------|--------------------|
| Fault | | | | | | |
| A | 1·2 | 6·5 | 9·5 | 13 | 8 | No output |
| B | 0 | 0 | 0 | 0 | 0 | No output |
| C | 2·1 | 22·5 | 23·9 | 23·2 | 23·2 | No output |
| D | 0·8 | 11·5 | 12 | 11·7 | 11·7 | Waveform as Fig. 4.28 |
| E | 1·0 | 11·7 | 12·5 | 11·7 | 11·7 | Waveform as Fig. 4.29 |
| F | 0 | 22·3 | 23·9 | 23·2 | 23·2 | No output |

# 5 Oscillator and Time Base Circuits

## 5.1 Principles of Oscillators

An oscillator is any device or circuit that produces an output which varies its amplitude with time. The output may be sinusoidal, square, pulse, triangular, or sawtooth as shown in Fig. 5.1.

These circuits are used in all types of electronic equipment, from radio and tv transmitters and receivers, computers, oscilloscopes, signal generators, to digital frequency meters.

Oscillators can be constructed using components that exhibit a negative resistance characteristic such as the unijunction transistor and the tunnel diode. The operation of these will be discussed later. However, a large majority of circuits are based round an amplifier with a positive feedback loop. When a portion of an amplifier's output is fed back in phase with its input, the effective input is increased and so is its overall gain.

For positive feedback

$$A_c = \frac{A_o}{1 - \beta A_o}$$

So, if the loop gain $\beta A_o$ approaches unity

$$A_c \rightarrow \frac{A_o}{0} \rightarrow \infty$$

The gain with positive feedback thus approaches infinity. Such high gains will result in oscillations, the frequency of which must be controlled by a frequency-determining network.

The requirements for a circuit to produce oscillations are
(a) Amplification
(b) A positive feedback loop
(c) Some network to control the frequency
(d) A source of power.

This is shown in block form in Fig. 5.2. The network that determines the operating frequency may be part of the feedback circuit or external to it. Typical circuits are combinations of L and C, or R

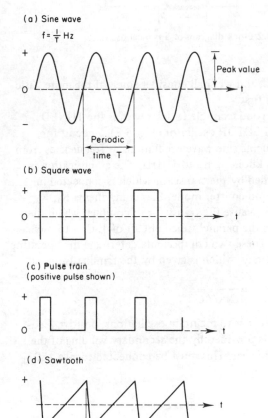

**Fig. 5.1** Typical oscillator output signals

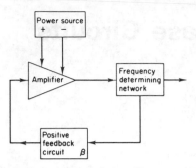

Fig. 5.2 Block diagram of a typical oscillator

and C, and we shall consider some of the more basic types first.

A good example to start with is the TUNED COLLECTOR oscillator (Fig. 5.3), a circuit for producing sine wave oscillations at frequencies from a few kilohertz up to 1 MHz. The amplification is provided by the transistor which is connected in common emitter mode. Bias components $R_1$, $R_2$ and $R_3$ cause the transistor to conduct, which then forces the parallel tuned circuit of $L_1C_1$ into oscillation. These two components determine the operating frequency, which is given by the familiar formula

$$f_0 = \frac{1}{2\pi\sqrt{LC}}$$

In order to maintain oscillations, positive feedback is provided by the secondary winding of the transformer. This must be connected to give 180° phase shift.

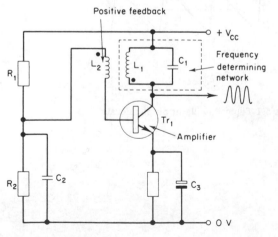

Fig. 5.3 Tuned collector oscillator

Another method of providing positive feedback in a single stage amplifier, without using a transformer, is to use a PHASE SHIFTING network from the collector to the base. A typical example of a phase shift oscillator is shown in Fig. 5.4. Each CR network from the collector to the base provides 60° of phase shift so that the resultant feedback is positive. This oscillator is particularly useful for generating sine waves of fixed frequency.

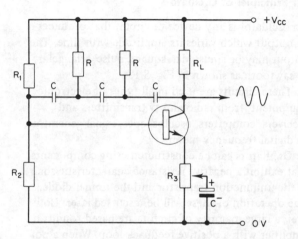

Fig. 5.4 Phase shift oscillator

Another circuit which depends upon phase shift is the Wien bridge oscillator shown in schematic form in Fig. 5.5. This is the standard circuit used in sine wave generators over the range of 1 Hz up to 1 MHz since it can easily be made continuously variable over that range. It has excellent stability and produces an output with low harmonic distortion. An example of this circuit is discussed later.

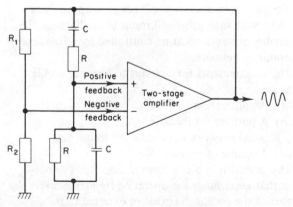

Fig. 5.5 Block diagram of Wien bridge oscillator

Square, pulse and sawtooth oscillators usually have an operating frequency that is determined by the charge and discharge times of a capacitor. Such circuits are used as clock pulse generators in digital circuits, time base generators in oscilloscopes, and pulse generators in radar.

It would not serve a useful purpose to detail further all the many forms of oscillators that are in current use. Other texts should be consulted for circuits of say the Hartley, Colpitts, Clapp, and so on. What we are concerned with are the important features, or characteristics, of the circuit, and possible fault conditions. Three of the main characteristics are

(1) Operating frequency
(2) Output amplitude
(3) Frequency stability.

Depending on the type of circuit however, other important characteristics will be

For *sine wave* oscillators

(1) The harmonic distortion (the purity of the sine wave).

For *square wave* and *pulse* oscillators

(1) The mark-to-space ratio, and width.
(2) The rise and fall times.
(3) Any overshoot or sag.

For *sawtooth* oscillators

(1) The linearity.
(2) The flyback time.

## 5.2 Measurement of Frequency

The measurement of frequency and output amplitude can be made using any standard method depending upon the required accuracy. An oscilloscope is perhaps one obvious choice, but even when properly calibrated this only gives an accuracy of about ±3% for both time and amplitude. This may be adequate for a large majority of cases, such as for example the measurement of the frequency of the bias oscillator in an audio tape recorder. When greater accuracy is required the unknown frequency must be measured by comparing it with a standard oscillator of known frequency. Using a CRO to obtain Lissajous figures is one well known method. More commonly the preferred method now is to measure the frequency by a digital frequency meter (Fig. 5.6). These instruments, fitted with an internal,

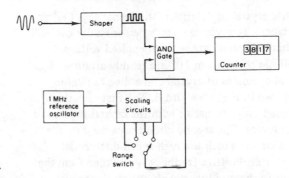

**Fig. 5.6** Block diagram of one form of frequency counter

highly stable and accurate oscillator, display the unknown frequency on a five digit or more in-line display. The accuracy of such instruments, dependent upon the accuracy of the internal oscillator, can be ±0·01% or better. The instrument counts the number of oscillations of the input frequency that occur within a time period determined by the internal oscillator and the display switch. For example suppose a gating time period of 1 millisec is selected, and the total count displayed is 1000, then the frequency of the input is 1 MHz.

## 5.3 Frequency Stability

The stability of the output frequency is very important in many applications. Various factors can cause the frequency of an oscillator to drift from the preset value. These include

(*a*) Changes in power supply voltage levels.
(*b*) Changes in active component parameter — transistor current gains, etc.
(*c*) Changes in load.
(*d*) Variations in components that determine the frequency.

The effect of the first three factors can be minimized by using stabilized power units, and a buffer amplifier between the oscillator and load. The biggest cause of instability will come from changes in the components that make up the frequency-determining network. Obviously components with good long-term stability, and very low temperature coefficient, should be used, and quite often these components are housed in a temperature-controlled enclosure. To achieve the highest possible stability, the designer must resort to using a piezo-electric crystal to determine the frequency. Certain substances, typically quartz, when specially cut, will mechanically resonate with an applied voltage. Stabilities of 1 part in $10^8$ are readily attained. Two typical examples of crystal controlled oscillators are shown in Fig. 5.7A and B. The first circuit is a modified Colpitts circuit with the crystal in place of the inductor. The second circuit uses the 710 integrated circuit which is a high-speed differential comparator. Positive feedback is provided from the output to the non-inverting input by the crystal. The d.c. operating point is fixed by $R_3$ and $R_2$, and $C_1$ decouples $R_2$, thus removing negative feedback at the oscillation frequency.

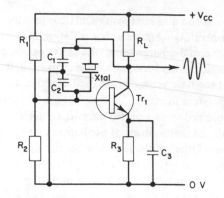

**Fig. 5.7A** Crystal-controlled Colpitts oscillator

## 5.4 Harmonic Distortion

This results from non-linearity or excessive gain in the amplifier circuit. With most oscillators the gain of the amplifier must be controlled to a value that just maintains the losses in the rest of the circuit. In Fig. 5.4 for example, the potentiometer can be adjusted to give more or less negative feedback. If the amplifier gain is too high then the amplitude of the oscillations builds up and the output distorts.

   Distortion is best measured by a distortion-measuring set, but assuming this is not available, another method can be to use a narrow, band stop filter. The oscillator output is passed through this filter, which must have very high attenuation at the oscillator's frequency, but which passes all harmonics. The resulting output from the filter can be measured with a true r.m.s. reading meter, and this can be used to calculate the percentage harmonic content of the oscillator's waveform. A twin T-filter is a suitable type.

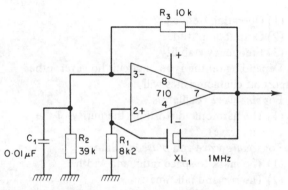

**Fig. 5.7B** 1 MHz crystal-controlled oscillator using a 710 comparator

(a) Single pulse

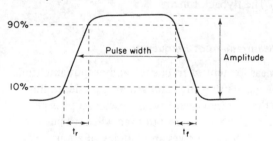

## 5.5 Square and Pulse Waveforms

Square wave and pulse waveform is shown in Fig. 5.8, and in some cases the same instrument is used to give either square or pulse output. Both waveforms are called rectangular, a square wave being a special case of a rectangular wave with a mark-to-space ratio of one.

   A typical pulse waveform is shown in Fig. 5.8, in which the various characteristics are defined. Note that the rise time of the leading edge is measured as

(b) Pulse train

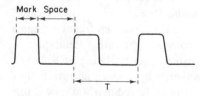

**Fig. 5.8** Pulse wave form
In(b), pulse repetition frequency = $\frac{1}{T}$ pulses per sec (Hz)

the time taken for the signal level to change from 10% to 90% of its full amplitude.

Pulse measurements are made using a wideband oscilloscope, and usually an external synchronizing signal must be provided in order that the trace displays the leading edge of the waveform. With rise time measurements, the rise time of the measuring instrument itself cannot be ignored, and nor must the measuring leads present a relatively high stray capacitance at the oscillator. An attenuating probe must be used to couple the signal into the Y-amplifier of the oscilloscope.

The rise time measured on the screen is related to the circuit waveform rise time and the oscilloscope's rise time by the formula

$$t_r = \sqrt{(t_m^2 - t_y^2)}$$

where $t_r$ is actual rise time

$t_m$ is measured rise time

$t_y$ is rise time of oscilloscope Y-channel.

An example will illustrate the effect of the oscilloscope's rise time. Suppose $t_m$ is 20 nanosecs and $T_y$ 15 ns. Then

$$t_r = \sqrt{(400 - 225)} = \sqrt{175} \simeq 13 \text{ ns}$$

Wherever the measured value approaches the oscilloscope's rise time this formula must be applied.

Square waves can be generated from relaxation oscillators such as the astable multivibrator or by passing the output of a sine wave oscillator through a squaring circuit such as a Schmitt trigger. This latter method is commonly used in general purpose laboratory instruments; a typical block diagram is shown in Fig. 5.9.

## 5.6 Sawtooth and Ramp Circuits

The majority of these circuits are those that produce a waveform which rises steadily with time up to a required amplitude, and then returns rapidly to the point from which the output can again rise. The linear rise is usually called the *sweep*, and the rapid return the *flyback*. These circuits find their main use in sweeping the beam across the face of a cathode ray tube, in other words in oscilloscopes, television cameras and receivers, and radar displays. In this case they are usually referred to as timebase circuits. The same type of circuit can be found in digital voltmeters where they are referred to as ramp generators. Circuits like these can be either free running or triggered from an external oscillator, but the basis of all of them is a capacitor which is charged and then rapidly discharged.

Apart from the frequency, the important requirement for most circuits is good linearity. This means that the rate of change of the output with time must be uniform. In fact, even in general purpose oscilloscopes, the linearity deviation in the time base should be better than 1%. When a capacitor is charged from zero volts towards a voltage $V$ via a resistor R, the voltage across the capacitor rises exponentially according to the formula

$$v_c = V(1 - e^{-t/CR})$$

Thus to obtain reasonable linearity from a simple discharger circuit (Fig. 5.10), the maximum change in $v_c$ should not be more than 10% of the total voltage.

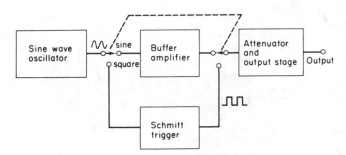

**Fig. 5.9** Block diagram of general purpose sine/square laboratory generator

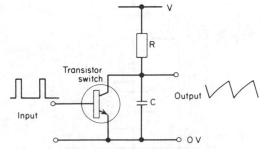

**Fig. 5.10** Simple discharger circuit for producing sawtooth wave forms

In this circuit the capacitor charges positively while the input to the transistor switch is zero. When a positive pulse is applied to the transistor switch it rapidly discharges the capacitor. The linearity from a simple circuit such as this is not usually sufficiently high, unless $V$ is very high. Various methods are used to overcome non-linearity, one being to charge the capacitor from a constant current source. A detailed example is dealt with later in this chapter.

## 5.7 Negative Resistance Oscillators

Strictly speaking, negative resistance oscillators and feedback oscillators are identical, since the latter can be regarded analytically as having introduced negative

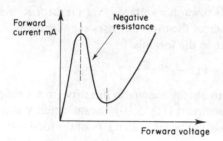

**Fig. 5.11** Tunnel diode characteristics

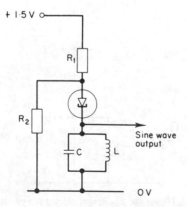

**Fig. 5.12** Typical tunnel diode oscillator

resistance into the circuit at the operating frequency. This negative resistance makes up the losses in the frequency network. However, it is best to classify negative resistance oscillators as those that use devices such as the tunnel diode, tetrode, unijunction transistor, i.e. devices that have an effective negative resistance region in their characteristics. A typical characteristic for a tunnel diode is shown in Fig. 5.11. The current first rises with forward voltage, then falls with increasing voltage, and finally rises again as the voltage is further increased. Placing a tunnel diode across a resonant circuit as shown in Fig. 5.12 provides an effective resistanceless tuned circuit which will then oscillate continuously. Very high frequency oscillators, up to as high as thousands of megahertz, are possible.

The unijunction transistor (UJT) is made of a bar of n-type material, (sometimes p) with ohmic contacts at each end and a p-type emitter junction formed near the centre (Fig. 5.13). The resistance of the bar is normally around 10 k$\Omega$, so when connected to a supply, with base 2 positive with respect to base 1, the bar acts as a potential divider, and a p.d. of $\eta V_{BB}$ appears between the emitter and $B_1$, where $V_{BB}$ is the voltage between $B_2$ and $B_1$ and $\eta$ is called the intrinsic stand-off ratio ($\eta$ is normally between 0·4 and 0·7).

When the emitter voltage is less than $\eta V_{BB}$ the emitter junction is reversed. When the applied emitter voltage exceeds $\eta V_{BB}$ by about 0·7 V, this voltage being called the peak point, the emitter becomes forward biased and injects holes into the $B_1$ region. Once this happens, the resistance between the emitter and $B_1$ falls to a low value. The action is regenerative.

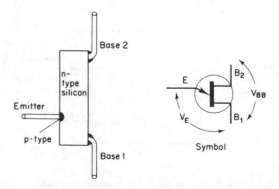

**Fig. 5.13** Construction of an n-bar unijunction transistor

A typical UJT relaxation oscillator circuit is shown in Fig. 5.14. Capacitor C charges via $R_1$ and eventually the emitter voltage exceeds the peak point of the UJT. The UJT conducts and discharges C rapidly via $R_3$. As C is discharged the current through the emitter falls, and when it falls below the minimum holding current then the UJT turns off. The capacitor can then charge again to repeat the process. In designing such circuits note that the value of $R_1$ should not be either too low, otherwise the UJT will not be able to switch off, or too high, otherwise the UJT may not receive sufficient emitter current to turn on. Typical values lie between 10 kΩ and 1 MΩ. UJT oscillators such as this are commonly used to trigger thyristor and triac circuits.

## 5.8 Fault Finding on Oscillators

Again because the types of circuit in common use are so various the fault finding procedure must be adjusted for the particular circuit and application. As with other parts of an instrument, a good understanding of the purpose and operation of the unit is essential. Wherever possible consult the maintenance manual before attempting measurements or adjustments in order to find if any special precautions or test instruments are necessary.

A number of oscillator units contain several blocks, such as attenuator, buffer amplifiers, modulators, etc., so a logical approach is essential to locate which block is non-functioning. For example, consider the block diagram of an amplitude modulated RF signal generator (Fig. 5.15) and suppose a fault exists such that there is

(a) An output from socket A, which is *continuous wave* with $SW_1$ on position 1, and modulated RF with $SW_1$ on position 2.
(b) No output from socket B.

The fault can only lie in the AF attenuator block.

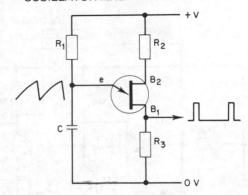

Fig. 5.14 Simple UJT relaxation oscillator

TABLE 5.1 Faults on an Oscillator Circuit

| FAULT | SYMPTOMS |
|---|---|
| Positive feedback loop open circuit | No output. D.C. bias levels correct. |
| Negative feedback loop open circuit | Output amplitude increased and wave shape distorted. |
| Open circuit component on one switch range | No output on that range only. |
| Bias component open circuit | No output all ranges. D.C. levels incorrect. |
| Switching transistor in ramp generator open circuit | Output from ramp permanently high. |
| Emitter follower with low or short circuit input impedance in ramp generator | Output from ramp reduced and non-linear. |
| Crystal open circuit in crystal controlled oscillator | Circuit will probably still oscillate but at a different frequency with poor stability. |

What would be the symptoms for

(a) An AF oscillator failure, or
(b) A modulator failure?

For individual oscillator circuits various faults can occur. Table 5.1 lists a few common symptoms.

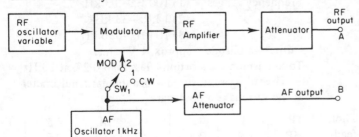

Fig. 5.15 Block diagram of an RF signal generator

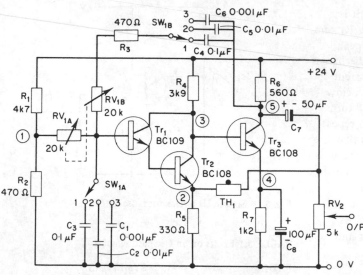

**Fig. 5.16** Wien bridge oscillator

## 5.9 Exercise: Wien Bridge Oscillator (Fig. 5.16)

The Wien bridge oscillator is a very popular circuit
for the generation of sine waves in the frequency
range 1 Hz to 1 MHz. The main reasons for this are

(a) The frequency depends upon the value of R
and C elements only, and high-grade R and C
components are more readily available than
inductors.

(b) The circuit has good stability and low distortion.

(c) The frequency can be made continuously
variable.

There is little to be gained from a detailed analysis of
the operation of the circuit as it is given in so many
textbooks. Briefly, it consists of a two-stage amplifier
with a positive feedback loop consisting of a series
RC network which develops a feedback signal across
a parallel RC network. The capacitors and resistors
are made equal in value, and the circuit oscillates at
a frequency when the phase shift from output to
input is zero degrees.

$$f_0 = \frac{1}{2\pi RC}$$

The attenuation in the positive feedback loop is
only a factor of 3, so the overall gain of the two-stage
amplifier has only to just exceed 3 to maintain
oscillations. A negative feedback loop is used to hold
the gain at this figure. In this case the negative feed-

back is from $Tr_3$ collector via thermistor to $Tr_2$
emitter. A thermistor is used to stabilize the output
amplitude. Its resistance will fall if the output
amplitude increases, and this increases the feedback
voltage, thus reducing the gain and automatically
reducing the output amplitude.

The frequency of Wien bridge oscillators can be
made variable over a wide range. In this case three
switched positions are shown which change the frequency by changing the value of the series and parallel
capacitors. The frequency is made continuous variable
by using a ganged potentiometer RV1A and RV1B
for the resistors. A small-value resistor (470 Ω) is left
in circuit when the potentiometer is at zero.

In the amplifier, $Tr_1$ is an emitter follower feeding
a common emitter amplifier $Tr_2$. In this way a high
input impedance is maintained. The output from
$Tr_2$ collector is directly coupled to $Tr_3$, another
common emitter amplifier.

The specification for the circuit is
Frequency ranges     (1) 100 Hz–3·3 kHz
                     (2) 1 kHz–33 kHz
                     (3) 10 kHz–330 kHz
Output amplitude   approx. 1 V r.m.s.
Total harmonic distortion   less than 0·2% at 1 kHz.

The voltages measured with a standard multimeter
at the various test points are

| TP | 1 | 2 | 3 | 4 | 5 |
|----|-----|-----|------|----|------|
| MR | 1·9 | 0·8 | 12·7 | 12 | 16·8 |

## Questions

(1) If the unit has failed such that no output can be obtained on switch position (3), which components are suspect, and with what type of fault?

(2) If the unit has failed such that no output can be obtained on any switch position, but the d.c. voltages appear normal, which components would be suspect? Give reasons.

(3) What would be the likely effect of (a) the thermistor going open circuit, (b) $C_4$ becoming open circuit?

(4) The unit fails to give an output on any range, the d.c. readings are as given below. State with reasons the faulty component.

| TP | 1 | 2 | 3 | 4 | 5 |
|----|---|---|---|---|---|
| MR | 0 | 0 | 16 | 15·2 | 15·3 |

(5) The unit fails to give an output on range (2) and the voltages measured with switch (2) made, and $RV_1$ at maximum, are

| TP | 1 | 2 | 3 | 4 | 5 |
|----|---|---|---|---|---|
| MR | 2·1 | 0 | 16 | 15·2 | 15·3 |

State with reasons the possible component fault.

(6) Write down the voltages you would expect to measure if $C_8$ became short circuit.

(7) The oscillator fails to give an output on any range. After studying the d.c. voltages, state which component is faulty and its type of fault.

| TP | 1 | 2 | 3 | 4 | 5 |
|----|---|---|---|---|---|
| MR | 2·1 | 0·8 | 13 | 12·5 | 24 |

---

### 5.10 Exercise: Blocking Oscillator Sawtooth Generator (Fig. 5.17)

This circuit is included because the blocking oscillator is such a useful circuit, and is quite often used in field time base oscillators in tv receivers.

The operation is as follows. As power is applied, $C_1$ charges via $R_1$ until the voltage at the transistor base reaches approximately 0·7 V. This will cause the transistor to conduct. Collector current flows and the collector voltage falls. The transformer coupling from collector to base is connected to give positive feedback; in other words the rise in collector current in the primary winding induces a voltage at the base that causes the transistor to conduct more. The transistor collector voltage rapidly falls to nearly zero and remains there while the collector current continues to increase. At some point, however, the collector current reaches a limiting value, usually because of the finite current gain of the transistor. When the primary current in the transformer stops changing there can no longer be an induced secondary voltage. The base voltage falls and turns off the transistor. Because of the positive feedback the transistor switches off rapidly and the base voltage goes negative. The negative step at the base equals the change in collector voltage divided by the turns ratio of the transformer. The transistor is held off while $C_1$ charges via $R_1$ towards +12 V. When

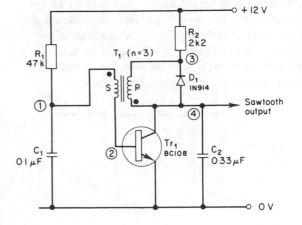

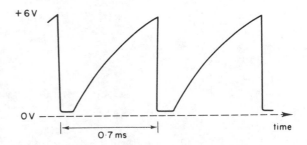

**Fig. 5.17** Blocking oscillator sawtooth generator

the voltage across $C_1$ reaches 0·7 V the transistor again turns on. Without $C_2$ in circuit the output waveform would be a series of short-duration negative pulses at a frequency mainly determined by $R_1$ and $C_1$.

The blocking oscillator is particularly useful in generating pulses of relatively high peak power in a train of low mean power. For example a transistor with a mean current rating of 100 mA can be used to give a peak pulse of 1 A. The average power dissipation is kept within the transistor's rating because the duty cycle is low. It is this property that is used to provide a sawtooth waveform by rapidly discharging the capacitor $C_2$ when the transistor conducts. While the transistor is off, $C_2$ charges via $R_2$ towards +12 V. Thus a sawtooth waveform is developed across $C_2$; the waveforms are shown in Fig. 5.17. Note that the linearity is not good.

When you build the circuit experiment with different values for $C_1$ and $C_2$, and note the effect on frequency and waveshape. The circuit should give a sawtooth of +6 V amplitude at a frequency of about 1·5 kHz. A small pulse transformer of turns ratio 3:1 was used, but if not available, a small audio driver transformer also works in the circuit.

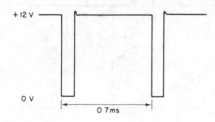

Fig. 5.18 Output wave form under a fault condition

## Questions

(1) The sawtooth generator fails and the voltage levels are

| TP | 1 | 2 | 3 | 4 |
|----|---|---|---|---|
| MR | 0 | 0 | +12 | +12 |

State, with reasons, the component or components that are faulty.

(2) The oscillator fails such that the output is as shown in Fig. 5.18. State which component is faulty and the type of fault.

(3) There are two component failures that could cause the oscillator to fail and give the following readings. State both.

| TP | 1 | 2 | 3 | 4 |
|----|---|---|---|---|
| MR | 0·7 | 0·7 | 0·1 | 0·1 |

(4) Write down the effect of the collector going open circuit.

(5) If $R_2$ became high in value, to say 4 k$\Omega$, what effect would this have on the circuit and output waveform?

(6) Which component is faulty if the oscillator fails and voltage readings are

| TP | 1 | 2 | 3 | 4 |
|----|---|---|---|---|
| MR | +9·7 | 0 | +12 | +12 |

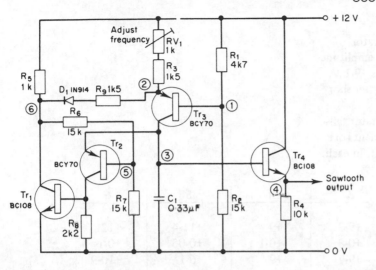

**Fig. 5.19** Free-running sawtooth oscillator with good linearity

## 5.11 Exercise: Free Running Sawtooth Oscillator (Fig. 5.19)

The previous circuit gives poor linearity, the reason being that the output capacitor is charged exponentially. A marked improvement in linearity results if the capacitor is charged from a constant current source. The circuit, which oscillates at a frequency of 500 Hz and gives an output of +6 V, can be considered as three blocks. $Tr_3$ is the constant current source, feeding a constant charging current of about 1 mA to $C_1$; $Tr_1$ and $Tr_2$ are connected as a triggered switch to rapidly discharge $C_1$; and $Tr_4$ is an emitter follower providing a low impedance output. To understand any circuit it is always best to consider it in blocks.

Taking first the constant current source; the base voltage of $Tr_3$ is fixed at a voltage of approximately +9 V by the potential divider $R_1$ and $R_2$. The emitter voltage will be at +9·7 V, which means that a voltage of 2·3 V is set up across $R_3$ and $RV_1$. If the total resistance of $R_3$ in series with $RV_1$ is set to 2k3, then the current flowing through the transistor will be 1 mA. This collector current will remain constant for a large change in collector emitter voltage. When $C_1$ is just discharged and the switching transistors $Tr_1$ and $Tr_2$ are off, this constant current will flow to charge $C_1$.

Now $Q = CV$ for a capacitor and since $Q = idt$,

$$dt = \frac{CV}{i}$$

where $i$ = constant current
$C$ = value of capacitor in farads
$V$ = ramp amplitude
$dt$ = ramp time.

With $i = 1$ mA, $C = 0.33\ \mu$F, and $V = 6$ V, then

$$dt = \frac{0.33 \times 10^{-6} \times 6}{1 \times 10^{-3}} = 1.98 \times 10^{-3} \text{ sec}$$

In other words for a 6 V amplitude the ramp time will be approximately 2 ms.

The emitter follower $Tr_4$ buffers the ramp signal to the output so that as little current as possible is diverted from the capacitor, otherwise non-linearity will result.

When the voltage across $C_1$ exceeds the base voltage of $Tr_2$ by 0·7 V, $Tr_2$ conducts. This in turn causes $Tr_1$ to conduct. Since there is positive feedback from $Tr_1$ collector to $Tr_2$ base, both transistors rapidly turn on. $C_1$ is discharged to a voltage level of nearly +0·7 V (equal to $Tr_1$ base emitter voltage). When $Tr_1$ collector voltage falls, $D_1$ conducts thus cutting off the constant current source. As soon as $C_1$ is discharged base current to $Tr_1$ ceases, and $Tr_1$ and $Tr_2$ rapidly turn off, thus allowing $C_1$ to again charge linearly.

To change the oscillator's frequency, switch in different values of capacitor. The output amplitude is controlled by the ratio of resistors $R_6$ and $R_7$, and these can be adjusted to change the amplitude.

**Questions**

(1) Which component is faulty if the generator frequency changes to nearly 1 kHz and its amplitude falls to 5 V? The d.c. level at test point 1 is +9·1 V.

(2) What would be the effect of a base emitter short on $Tr_4$?

(3) In each of the following cases the generator fails to produce an output. State which component (or components) is faulty and the type of fault. In each case give a supporting reason.

|  | 1 | 2 | 3 | 4 | 5 | 6 |
|---|---|---|---|---|---|---|
| Fault |  |  |  |  |  |  |
| A | +9·1 | +9·8 | +9·7 | +9 | +11·4 | +12 |
| B | +9·1 | +9·8 | +0·8 | +0·1 | +0·05 | +0·1 |
| C | +9·1 | +9·8 | +9·7 | +9 | +6·1 | +10·9 |
| D | +12 | +5·3 | +0·8 | +0·1 | +0·05 | +0·1 |
| E | +9·1 | +5·3 | +0·8 | +0·1 | +0·05 | +0·1 |

(4) The oscillator fails so that its output amplitude falls to less than 1 V. The voltages measured:

| TP | 1 | 2 | 3 | 4 | 5 | 6 |
|---|---|---|---|---|---|---|
| V | +9·1 | +9·8 | – | – | 0 | +12 |

State which component is faulty and the type of fault.

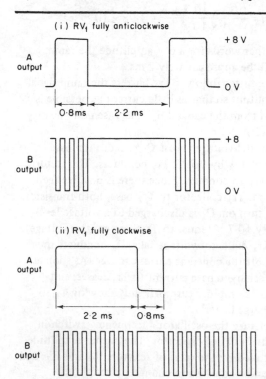

(i) RV₁ fully anticlockwise

A output
+8 V
0 V
0·8ms  2·2 ms

B output
+8
0 V

(ii) RV₁ fully clockwise

A output
2·2 ms  0·8ms

B output

**Fig. 5.22** Output wave forms from gated oscillator

Fig. 5.22 applies to the Exercise on the Gated UJT Pulse Generator on page 74. It has been placed on this page for convenience of space and layout.

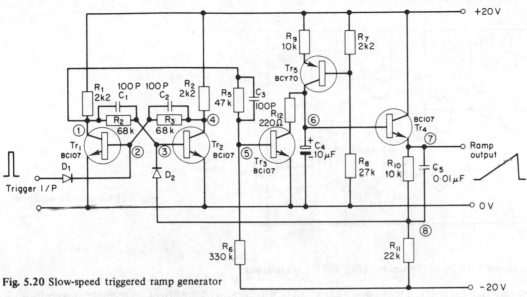

**Fig. 5.20** Slow-speed triggered ramp generator

### 5.12 Exercise: Slow Speed Ramp Generator (Fig. 5.20)

This circuit is designed to provide a single ramp output signal for one positive input trigger pulse. The ramp duration is approximately 1 sec and the amplitude is 10 V. As in the previous example the timing capacitor $C_4$ is charged from a constant current source $Tr_5$. A tantalum capacitor should be used.

$Tr_1$ and $Tr_2$ form a bistable circuit with $Tr_1$ normally off and $Tr_2$ on. Under these conditions the collector voltage of $Tr_1$ will be high and this foward biases $Tr_3$ via $R_5$. $Tr_3$ is the switching transistor which clamps the voltage across $C_4$ to nearly zero volts.

When a positive input pulse is applied to $Tr_1$ base, the bistable changes state with $Tr_1$ on and $Tr_2$ off. The forward bias to $Tr_3$ is switched off, and $C_4$ can now charge. The voltage across the capacitor rises positively. This linearly rising voltage is fed via an emitter follower $Tr_4$ to the output. An emitter follower is used to provide a low-output impedance

and to prevent the load from unduly affecting the linearity of the ramp. As the output rises so does the voltage at the junction of $R_{10}$ and $R_{11}$. When the voltage at this point reaches approximately +1·0 V, the diode $D_2$ conducts and switches on $Tr_2$ to reset the bistable. $Tr_3$ is switched on and $C_4$ is rapidly discharged.

A unit such as this can be useful for checking the characteristics of devices when used in conjunction with an XY plotter.

### Questions

(1) How could the circuit be modified to be free running?

(2) What is the purpose of $C_5$?

(3) In each of the following fault conditions the circuit fails to operate correctly when an input trigger is supplied. State which component (or components) is at fault and give reasons for your choice. The voltages were measured with respect to 0 V using a standard multimeter.

| Fault | 1 | 2 | 3 | 4 | 5 | 6 | 7 | 8 |
|---|---|---|---|---|---|---|---|---|
| A | +0·1 | +0·7 | +0·05 | +19·6 | −1·6 | +19·5 | +18·9 | +6·4 |
| B | +19·6 | 0 | +0·7 | +0·1 | −8 | +19·5 | +18·9 | +1·4 |
| C | +0·1 | +0·7 | +0·05 | +19·6 | −1·6 | 0 | −0·6 | −6·7 |
| D | +19·6 | +0·2 | +0·75 | +0·75 | +0·7 | 0 | −0·6 | −6·7 |
| E | +19 | 0 | +0·7 | +0·1 | +0·7 | +0·05 | −0·6 | −6·7 |
| F | +19 | 0 | +0·7 | +0·1 | 0 | +19·6 | +18·9 | +1·4 |
| G | +0·1 | +0·7 | +0·05 | +19·6 | +0·7 | +0·7 | 0 | −6·2 |
| H | +0·1 | +0·7 | +0·05 | +19·6 | −1·6 | +19·6 | +19 | −19·6 |

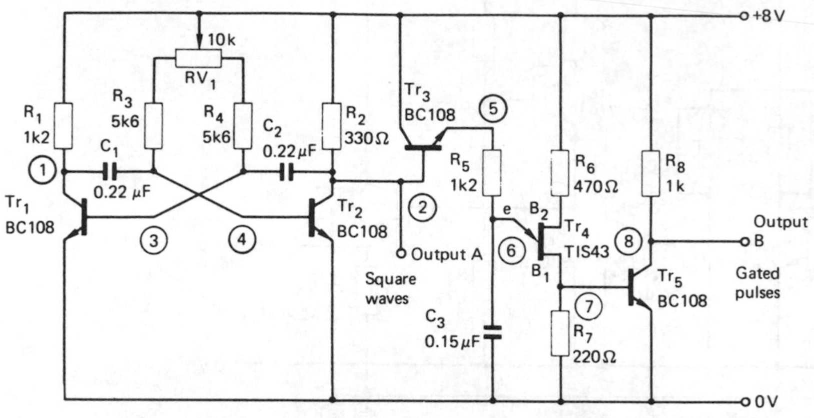

**Fig. 5.21** Gated pulse generator

## 5.13 Exercise: Gated UJT Pulse Generator (Fig. 5.21)

As stated previously the unijunction transistor makes an excellent relaxation oscillator. In this circuit the emitter supply of the unijunction is fed via a transistor switch from an astable multivibrator. The low frequency waveform from the astable is used to gate the unijunction, so that a burst of pulses appear at the output.

$Tr_1$ and $Tr_2$ from an astable multivibrator, oscillating at a frequency of approximately 300 Hz. The mark-to-space ratio of the waveform at $Tr_2$ collector can be varied over a fairly wide range by the potentiometer $RV_1$. When $Tr_2$ switches off, $Tr_3$ conducts and so its emitter rises to nearly +8 V. $C_3$ then charges rapidly via $R_5$, the unijunction triggers discharging $C_3$ and a positive pulse appears across $R_7$. This causes $Tr_5$ to conduct, and a pulse is generated at the output. While $Tr_2$ remains off, the unijunction circuit continues to oscillate producing pulses at the output. The number of pulses appearing at the output is determined by the mark/space ratio of the astable multivibrator. This is shown in Fig. 5.22* where it can be seen that few pulses are produced for a low mark/space ratio, and ten pulses are produced for the maximum mark/space ratio. The operation can be modified by changing values of $R_3$, $R_4$, or $C_3$.

### Questions

(1) What would be the effect on the output waveforms if $Tr_3$ developed a base emitter open circuit?

(2) The circuit develops a fault so that the output waveform at B is the inversion of the waveform at A. Which component is faulty?

(3) What would be the effect on the circuit operation if $Tr_1$ developed a base emitter short?

(4) Which component (or components) is at fault if a continuous train of negative pulses of 0·1 ms duration appears at output B. The voltages measured at the test points were

| TP | 1 | 2 | 3 | 4 |
|----|-----|-----|-----|-----|
| MR | 0·1 | 7·9 | 0·7 | 0 |

(5) State, for the following fault conditions, the component that is faulty and its type of fault. In each case the fault symptoms are that there is no output waveform at (B) while a 300 Hz square wave is present at (A).

| Fault | 6 | 7 | 8 |
|-------|------|------|---|
| 1 | 0 | 0·3 | 8 |
| 2 | 0·32 | 0·32 | 8 |
| 3 | 5·4 | 0·3 | 8 |
| 4 | 3·6 | 0 | 8 |
| 5 | 0·4 | 0 | 8 |

*Fig. 5.22 appears on page 72.

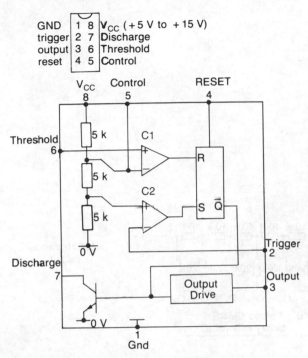

GND | 1 | 8 | V_CC (+5 V to +15 V)
trigger | 2 | 7 | Discharge
output | 3 | 6 | Threshold
reset | 4 | 5 | Control

**Fig. 5.23** Internal parts of the 555

### 5.14 Exercise: A 555 monostable circuit

ICs such as the 555 timer are very useful in waveform generating applications. The internal arrangement of a 555 consists of a precision resistor network, two fast comparators, a bistable, a switching (discharge) transistor and an output drive circuit. The output can source or sink up to 200 mA of load current and switching speeds are 100 nsec. In the monostable mode two external components, a timing resistor and capacitor, are required to set the pulse width, see Fig. 5.23. When power is first applied, without a trigger input, the $\bar{Q}$ output of the bistable is high forcing the discharge transistor to be on, and at the same time holding the output low. The three internal 5 kΩ resistors form a simple divider chain setting up voltages as follows:

$\frac{2}{3}V_{CC}$ on the inverting input of comparator 1 and $\frac{1}{3}V_{CC}$ on the non-inverting input of comparator 2.

When a negative trigger pulse is applied to pin 2 the output of comparator 2 immediately goes high and this sets the bistable. $\bar{Q}$ output goes low, the discharge transistor turns off and the output rises

high to about 1 V less than $V_{CC}$. The external timing capacitor $C_T$ can now charge via $R_A$ and the voltage across it rises exponentially towards $V_{CC}$ with a time constant set by the product of $R_A$ and $C_T$. When this voltage just exceeds $\frac{2}{3}V_{CC}$ the output of comparator 1 goes high and resets the bistable. The discharge transistor switches on and rapidly discharges $C_T$ to 0 V. At the same time the output on pin 3 returns to zero volts. The width of the pulse generated at the output is given by:

$$T = 1 \cdot 1 \ C_T R_A$$

An example for a 5 sec delay circuit using a 555 timer is shown in Fig. 5.24. In this circuit the timing is initiated by a momentary action push-switch $SW_1$. The state of the output is indicated by a simple LED circuit.

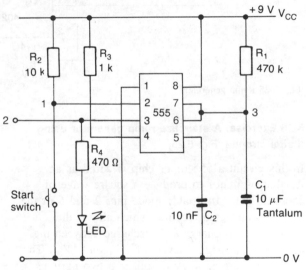

**Fig. 5.24** 555 monostable circuit

### Questions

(1) Sketch the waveforms that you would normally expect to find at the three test points. Ensure that your waveforms are correctly time related.
(2) State as fully as possible the symptoms for the following faults
    (a) $C_1$ short circuit
    (b) $C_1$ open circuit
    (c) $R_1$ open circuit
    (d) An open circuit track to pin 2
    (e) A short circuit between pins 1 and 2.

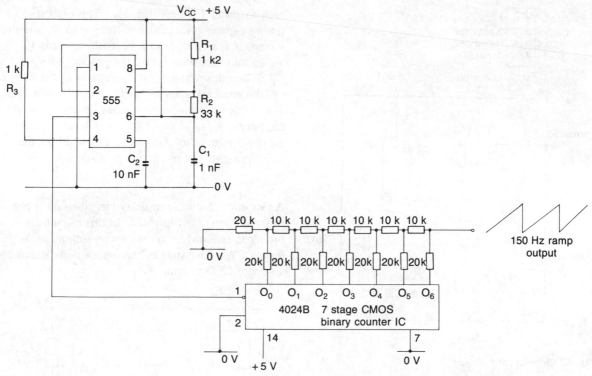

**Fig. 5.25** Ramp generator

### 5.15 Exercise: A staircase ramp generator using digital circuits (Fig. 5.25)

In this circuit a 555 timer chip is wired as an Astable oscillator to produce a square wave of about 20 kHz. In astable mode pins 2 and 6 of the 555 are connected together which allows the capacitor C to charge and discharge between the threshold and trigger levels of $\frac{2}{3}V_{CC}$ and $\frac{1}{3}V_{CC}$. The timing resistor network is split into two parts as shown in the diagram. The output waveform will not be symmetrical but with $R_2$ made much larger than $R_1$ an approximation to a square wave results at the output. The frequency is given by the formula:

$$f = \frac{1}{t_1 + t_2}$$

where $t_1 = 0\cdot7(R_1 + R_2)C_1$ and $t_2 = 0\cdot7R_2C_1$, and when $R_2 \gg R_1$, as in this case

$$f \approx \frac{1}{1\cdot4R_2C_1}$$

The output from the oscillator is connected to the

clock pulse input of a CMOS 7 stage binary counter (4024B) which will increment every time the oscillator output makes a transition from a high to a low, i.e. on the trailing edge of the 555's output waveform. The seven outputs of the 4024B, $O_0$, $O_1$, $O_2$ up to $O_6$, are connected to a simple digital to analogue converter constructed from two values of resistor, 20 k$\Omega$ and 10 k$\Omega$. This R-2R ladder, as it is called, is a standard part of most DACs. Since the counter is being continuously advanced the output of the ladder network will be a ramp consisting of 127 steps. This ramp will therefore have a frequency of:

$$f_R = \frac{f_{0(clock)}}{2^n} \quad \text{where } n = \text{no. of bits}$$

therefore $f_R \simeq 150$ Hz

### Questions

(1) Calculate the exact value of the output frequency assuming all timing components are as stated.

(2) The circuit fails to deliver a ramp output. Describe the next test to be made to locate the fault.

(3) Explain how the 7 stage binary counter could be tested separately from the 555 oscillator.

(4) It is required to slow the operation of the circuit down considerably. Explain how this can be achieved without desoldering any components.

(5) The output becomes a square wave with a frequency of approximately 150 Hz. Explain, with reasons, the likely fault in this circuit.

# 6   Pulse and Waveform Shaping Circuits

## 6.1 Introduction

Much of modern electronics is concerned with shaping and modifying pulses and other signals. The main pulse-forming and waveform-shaping networks can be grouped into

(1) Linear passive circuits: those made up of R, C and L elements.

(2) Non-linear passive circuits: diode clippers and restorers.

(3) Active circuits: those that use transistor switches such as the Schmitt trigger and monostable.

Some of the more common examples will be dealt with in this chapter. Such circuits will be found in colour tv receivers, radar sets, and in fact in nearly all electronic equipment.

## 6.2 Linear Passive Circuits — the Integrator and Differentiator

Circuits that contain only R, C or L components are termed linear, since they do not affect the shape of a sine wave input signal. They only produce attenuation and phase shift for a pure sine wave. However, they greatly modify other wave shapes.

The INTEGRATOR, often called a low pass filter, is shown in Fig. 6.1 together with output signals for typical inputs. Since the reactance of the capacitor falls with increasing frequency, this circuit removes the high-frequency components from a pulse waveform. When a step input is applied the voltage across the capacitor cannot change instantaneously. It rises exponentially according to the formula

$$v_c = V(1 - e^{-t/CR})$$

Now $CR,$ the product of capacitance in farads and resistance in ohms, is called the TIME CONSTANT of the circuit. In one time constant the voltage across the capacitor changes by about 63%. Note that it takes nearly 4·5 time constants for the voltage across the capacitor to equal $V.$

The effect of an integrator on pulses which have a long width in comparison to the "integrator's" time constant is to degrade the rise and fall times. If, however, the pulse is short in comparison to the "integrator's" time constant, then the capacitor will not have sufficient time to charge completely, and the output will appear triangular. Circuits such as these are often used to provide short time delays. An example is shown in Fig. 6.2.

The DIFFERENTIATOR, basically a high pass filter, allows high frequencies to pass, but attenuates the low frequencies. The circuit together with outputs for various pulse inputs is shown in Fig. 6.3. When a step waveform is applied, and assuming that C is uncharged, then the voltage across the capacitor cannot change instantaneously. The voltage across a capacitor can only change when it acquires some charge, and this naturally takes time. So the output must rise to the same value as the input. As C charges the voltage across it increases, and the output voltage across R falls exponentially:

$$v_r = V(e^{-t/CR})$$

The result is that a "spike" equal to the change of state at the input is generated.

If an input pulse that is long compared to the differentiator's time constant is applied, then the output must go negative on the trailing edge. This occurs because the capacitor, already charged by the leading edge of the pulse, cannot change its voltage instantaneously when the trailing edge arrives. The left-hand plate of the capacitor will be at +V and the right hand plate at 0 V. When the input changes abruptly from +V to 0 the output must now change from 0 to −V.

Differentiators are often used to convert pulses of one polarity into "spikes" of the opposite polarity. These spikes can then be used to trigger other circuits.

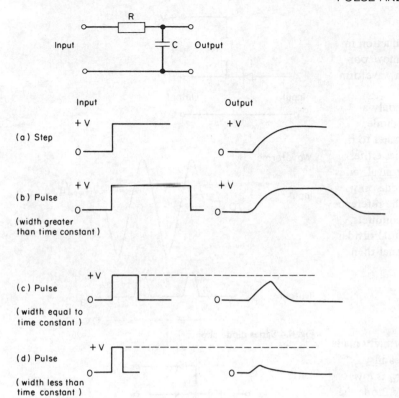

**Fig. 6.1** An integrator

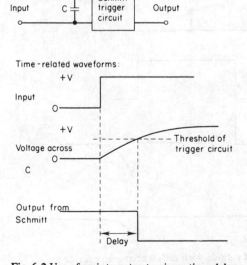

**Fig. 6.2** Use of an integrator to give a time delay

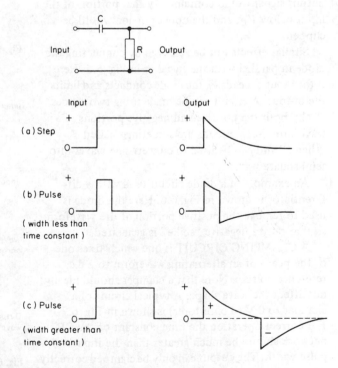

**Fig. 6.3** A differentiator

### 6.3 Diode Waveform Shapers

Diodes, because of their property of conduction in one direction only, are widely used to remove portions of a signal (clipping) and to clamp a waveform to a reference level (restoring).

CLIPPING CIRCUITS are used where only a portion of the input signal is required. A diode, either in series with the signal path or parallel to it, is used to remove part of the signal that lies either above or below a d.c. reference level. A typical series diode clipper is shown in Fig. 6.4. The diode only conducts when the input signal exceeds the reference bias level $V_B$. Naturally when the diode conducts, its forward slope resistance forms a potential divider with R to the input signal. The output signal then will be attenuated slightly, i.e.

$$V_o = V_i \left( \frac{R}{R + r_d} \right)$$

Therefore, to minimize errors, $R$ is normally made much greater than $r_d$. In practice this is readily achieved since the diode slope resistance $r_d$ is low, typically less than 100 $\Omega$. By reversing the diode the output signal would contain only that portion of the input below $V_B$, and the upper portion would be clipped.

Similar effects can be achieved by connecting the diode in parallel with the signal as in Fig. 6.5. Here if the input exceeds $V_B$ the diode conducts and limits the output. A circuit can be made using two diodes to clip both the positive and negative portions of a waveform. Such circuits are sometimes called "slicers" and can be used to convert sine waves into semi-square waves.

An example of a diode circuit used with a differentiator is shown in Fig. 6.6. Here the diode is used to remove the positive portion of the waveform so that only a negative "spike" is generated.

A CLAMPING CIRCUIT is one which fixes one of the peaks of an alternating waveform to a d.c. reference voltage. Note that a clamper should ideally not affect the wave shape. A typical circuit of a clamper to 0 V (d.c. restorer) is shown in Fig. 6.7. For correct operation the time constant of the CR network should be much greater than the input pulse width. The output can only be clamped correctly when the input consists of a train of pulses as shown.

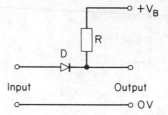

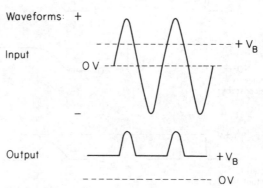

**Fig. 6.4** Series diode clipper

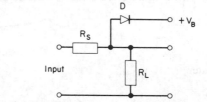

**Fig. 6.5** Shunt diode clipper

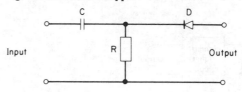

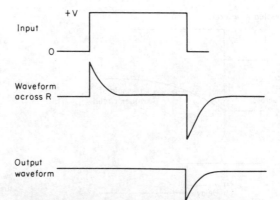

**Fig. 6.6** Use of a diode with differentiator to provide a negative spike from a positive pulse input

The capacitor charges when the input waveform goes negative since the diode conducts heavily. After a few cycles of the input waveform the capacitor becomes charged to the peak value of the input, and therefore shifts the mean output level positive. By reversing the diode the positive peaks could be clamped to 0 V.

A good example of the use of diodes is in a field-synchronizing-pulse generator used in a monochrome tv receiver. An example is shown in Fig. 6.8 together with the waveforms. The diode $D_1$ is cut off by each negative synch pulse and this allows $C_2$ to discharge through $R_2$. When the synch pulse returns positive, $D_1$ conducts to recharge $C_2$ rapidly. The time constant of $C_2R_2$ is such that only a small negative signal is generated at $D_2$ anode by the narrow line synch pulses. However, when the broad field synch pulses are present, a large negative signal is generated. A diode clipping circuit $D_2R_3R_4$ removes any line synch information, and the signal is differentiated by $C_3R_5$ to produce a series of sharp field pulses at the output.

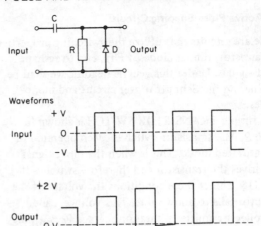

Fig. 6.7 Basic d.c. restorer circuit

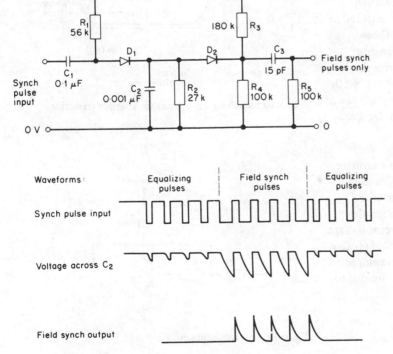

Fig. 6.8 Synch pulse separator circuit

## 6.4 Active Pulse Shaping Circuits

These are circuits that utilize the switching properties of transistor, tunnel diodes, FETs, etc. to reshape input signals. Under this general heading we shall be considering the Schmitt trigger circuit and mono-stables.

A simple TRANSISTOR SWITCH is shown in Fig. 6.9. The transistor switch is often operated in the saturated mode. This is when the input signal overdrives the transistor and therefore switches it hard ON. Under these conditions the voltage at the collector falls to a low value. This voltage, called the collector emitter saturation voltage $V_{CE(sat)}$, may have a value from 0·1 V to 0·6 V depending upon the type of transistor in circuit. The transistor is OFF when no input signal is present, or when the input is lower than the required base emitter voltage. Under these conditions the collector voltage is high, at $V_{cc}$, and the only current flowing through the collector load is the transistor's leakage current. With modern silicon transistors this is very small, and in most cases can be neglected.

When switched to the ON position there is a finite time before the output falls to $V_{CE(sat)}$; this is the transition time for charge carriers to move through the transistor. When switched OFF, however, there is a finite delay time before the collector current ceases, because minority charge carriers are stored in the base region of a saturated transistor. These charges have to be removed before the transistor switches to the OFF position. The waveforms showing transition and storage delays are shown in Fig. 6.10.

Often switching speeds are improved by the use of

(a) Speed-up capacitors: capacitors in parallel with drive resistors.

(b) An anti-saturation modification.

The latter is a circuit that prevents the transistor from saturating (Fig. 6.11). The diode $D_1$ is used to hold the collector voltage at a level that just prevents saturation. When an input is applied the collector voltage falls, but when the collector voltage is lower than that of the base, the series diode (often germanium) conducts and diverts the excess input current into the collector. You may find circuits in use with $R_2$ replaced by a silicon diode.

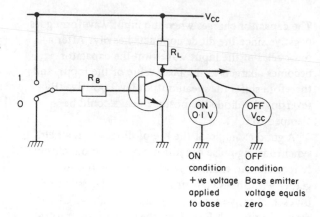

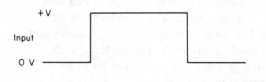

Fig. 6.9 Simple transistor switch

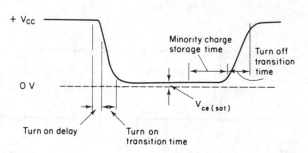

Fig. 6.10 Switching wave forms for a simple transistor switch

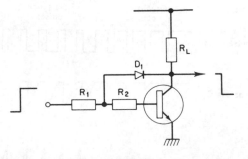

Fig. 6.11 Anti-saturation circuit

## 6.5 The Schmitt Trigger Circuit

This circuit, shown in Fig. 6.12, is used for level detection, reshaping pulses with poor edges, and squaring sine wave signals. The Schmitt is basically a snap-action switch that changes state at a specific trip point. Consider conditions when the input to $Tr_1$ base is at zero. $Tr_2$ is conducting since it has forward bias provided by the potential divider $R_2$, $R_3$ and $R_4$. The voltage at $Tr_2$ base is approximately

$$V_{B2} \simeq \frac{V_{CC}R_4}{R_2 + R_3 + R_4}$$

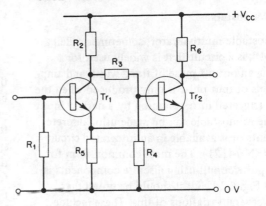

**Fig. 6.12** Basic Schmitt trigger circuit

The voltage at $Tr_1$ and $Tr_2$ emitter will be 0·7 V less than $V_{B2}$ and this positive voltage reverse biases $Tr_1$, thereby holding it off. The current flowing through $Tr_2$ is determined by

$$I_{C2} \simeq \frac{(V_{B2} - 0·7)}{R_5(k\Omega)} \text{mA}$$

and the collector voltage of $Tr_2$ will be

$$V_{C2} = V_{CC} - I_{C2}R_6$$

Usually the circuit is designed so that $Tr_2$ is not saturated, thus allowing faster switching speed.

When the input voltage is increased so that it nearly equals the voltage on $Tr_2$ base, then $Tr_1$ starts to conduct, its collector voltage falls, and $Tr_2$ starts to turn off. Because of the positive feedback between the emitters, $Tr_1$ rapidly turns on and $Tr_2$ off. The output voltage rises to $+V_{CC}$.

A special feature of the Schmitt is that the circuit does not switch back as soon as the input signal is reduced just below the threshold level or trip point, but at a much lower level. The circuit possesses hysteresis or backlash, and this is very useful in eliminating

noise superimposed on the input signal. A detailed analysis of the circuit is outside the scope of this book. But the reason for hysteresis can be seen by considering that the circuit changes state at the point when the two base voltages are equal. When the threshold level or upper trip point is passed, $Tr_1$ switches on and its collector voltage falls. This means that $Tr_2$ base voltage also falls, so in order for the circuit to switch back to its original state the input voltage to $Tr_1$ base must be reduced to a value equal to the lower voltage on $Tr_2$ base. The effect of the hysteresis of the circuit is shown in Fig. 6.13, where it can be seen that the output switches back only when the input is reduced below the lower trip point. A dual Schmitt trigger is available in TTL integrated circuit (SN 7413), and this is often used in interface circuits to improve the noise margin.

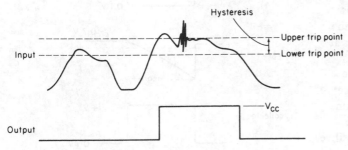

**Fig. 6.13** Typical wave forms showing the usefulness of the hysteresis in the circuit

## 6.6 The Monostable

The monostable multivibrator, sometimes called a "one shot", is a circuit that is widely used for generating an output pulse of fixed width and amplitude. This output pulse is only produced when the circuit is triggered into operation by a narrow input pulse. The monostable can be made using discrete components or is available in an integrated circuit package (SN 74121). The most common form for producing the circuit using discrete components is shown in Fig. 6.14, but it should be noted that there are several variations of this. These include emitter coupled and complementary types.

The basic circuit can be seen to consist of a two-stage amplifier with resistive coupling from output to input. As the name suggests, the circuit has one fixed stable state. This is with no input trigger pulse, when $Tr_2$ is ON and $Tr_1$ is OFF. $Tr_2$ conducts because it has forward bias provided by $R_t$. This resistor has a value low enough to provide sufficient base current to just drive $Tr_2$ into saturation. The collector voltage of $Tr_2$ will then be approximately 0·1 V, and this ensures that $Tr_1$ is held off in a non-conducting state. The circuit can be switched into a "quasi-stable" condition by applying a positive pulse to $Tr_1$ base. This need only be of short duration, as its purpose is merely to trigger the circuit into operation. The pulse causes $Tr_1$ to conduct and its collector voltage falls. This change in voltage is coupled via $C_T$ to $Tr_2$ base. Remember that the voltage across a capacitor cannot change instantaneously, so a change of voltage on one plate appears as an equal change on the opposite plate. The voltage on $Tr_2$ base therefore goes negative and this turns off $Tr_2$. The collector voltage rises towards $V_{cc}$ and, because of the positive feedback via $R_3$, $Tr_1$ is forced to conduct more. Very rapidly the circuit switches state so that $Tr_1$ is ON and $Tr_2$ is OFF, but this is not a permanently stable condition. The base of $Tr_2$ is negative, while $Tr_1$ collector is at approximately +0·1 V, so the capacitor $C_T$ now charges via $R_t$ and $Tr_1$ so that its right-hand plate moves from $-V_{cc}$ towards $+V_{cc}$. When this voltage reaches nearly +0·6 V, $Tr_2$ begins to conduct again, and the circuit rapidly switches back to its stable state. The output from $Tr_2$ collector is thus a positive pulse of amplitude approximately $V_{cc}$ and with a defined width. The width of the pulse is determined by the time constant $C_t R_t$ and is approximately

$$t_d \simeq 0.7\, C_t R_t$$

Various modifications can be made to improve the performance of the monostable, and circuits will be found in current use that use catching diodes, protection diodes, speed-up capacitors, etc. Most of these modifications are used to improve the operating speed of the circuit so that an output pulse with fast rise and fall times is achieved. A typical best figure for the rise time is 10 ns. The circuit shown in Fig. 6.15 includes catching diodes which operate to limit the output amplitude and improve speed. When $Tr_2$ switches off, its collector load has to charge any capacitive load which is formed by $C_L$ and stray circuit capacitance. Initially the rate of rise is fast but, without the diode, the total rise time would be relatively slow. With the diode in circuit, the change in voltage at the collector is limited and therefore the rise time is improved.

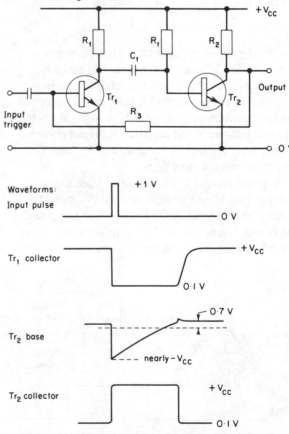

Fig. 6.14 Basic transistor monostable circuit

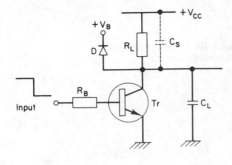

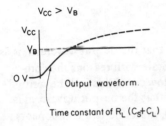

**Fig. 6.15** Circuit with catching diode

## 6.7 Fault Finding in Pulse and Waveform Shaping Circuits

Some of the faults that occur in pulse shaping and switching circuits are different in form from those in other units, such as amplifiers and power supplies. Quite often the signal is degraded in some way, so that the required wave shape is not produced. This may happen with or without a change in d.c. bias conditions. Locating such faults requires a good understanding of the circuit function. Another example of the special faults that can occur is in a switching circuit, such as a monostable, which produces output pulses or changes state when no input is present. This is called "spurious triggering" and it can be, in some cases, very difficult to locate the cause of the problem. The fault may lie in the circuit itself, but more usually will be caused by interference. This is "noise", either picked up on the input leads or travelling along the mains wires (mains borne). Such interference is often caused by rotating electrical machines or switching surges from heavy inductive loads situated near the electronic unit. In a very noisy industrial environment special care has to be taken in the design and installation of electronic units. This involves screens, mains filters, screened leads, and high noise immunity interface circuits. An example of the latter is discussed later.

When testing wave-shaping circuits ensure that the bandwidth of the oscilloscope that you use is sufficient for the measurement. It is good practice to use an attenuating probe so that the capacitive loading of the measuring leads are kept to a minimum.

With circuits that contain active switches together with linear and non-linear wave-shapers it's a good idea to follow a standard sequence in fault location:

(a) Measure the power supply lines with a multimeter. The voltages should be near the correct value and the ripple should be low.

(b) Check that the input signal is present. Often the input is generated from a transducer (photocell, thermistor) or a mechanical switch. Since the input-actuating device is usually positioned well away from the unit, it is much more prone to damage than internal components.

(c) Check that input leads and plug and socket connections are good. Ensure that any screen leads are properly earthed.

(d) If the actuating device is OK, cause the input signal to change state rapidly by operating the input device, or apply a suitable input from an oscillator. Then, by following the signal flow from input to output, check each stage until the signal is either degraded or lost. This will locate the faulty stage.

(e) The operation of individual transistor switching circuits can be checked without unsoldering the device. Remember that the circuit will be causing the transistor to be either ON or OFF. For a transistor that is ON, momentarily short the base to emitter. The transistor should turn OFF and its collector voltage rise to $V_{cc}$. If the circuit conditions are holding the transistor in the OFF state, check that it can be turned ON by applying a forward bias current. Use a resistor of about 10 kΩ temporarily connected from the supply rail to the base connection. The transistor should turn on and its collector emitter voltage should fall to a low value (typically 100 mV). See Fig. 6.16.

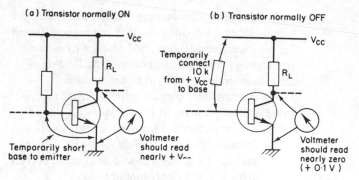

(a) Transistor normally ON
(b) Transistor normally OFF
Vcc
RL
Temporarily connect 10 k from + Vcc to base
Temporarily short base to emitter
Voltmeter should read nearly + Vcc
Voltmeter should read nearly zero (+ 0·1 V)

**Fig. 6.16** Rapid checking of transistor switches

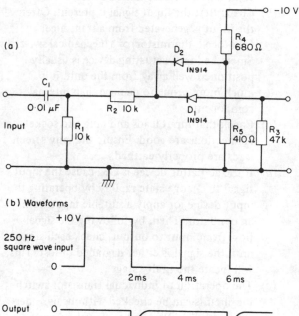

(a)
-10 V
R4 680 Ω
D2
IN914
C1
0·01 μF
R2 10 k
D1 IN914
Input
R1 10 k
R5 410 Ω
R3 47 k
(b) Waveforms
+10 V
250 Hz square wave input
0
2ms
4 ms
6 ms
Output
0
-4 V

**Fig. 6.17** Waveform shaping circuit

## 6.8 Exercise: Waveform Shaping Circuit (Fig. 6.17)

This circuit is designed to produce negative pulses of 4 V amplitude with a pulse width of about 0·2 ms from a 250 Hz 10 V amplitude square wave input.

$C_1$ and $R_1$ form a differentiator with a time constant of 0·1 ms. The waveform at the junction of $R_1$ and $R_2$ will therefore consist of positive and negative pulses. The series diode $D_1$ removes the positive signals, since it will only conduct on the negative spike. Diode $D_2$ is used as a negative clipper. It conducts when the signal at $R_2D_1$ junction goes more negative than the voltage set up by the potential divider $R_4$ and $R_5$. The time-related waveforms are shown in Fig. 6.17B.

### Questions

(1) The circuit fails so that the output waveform is as shown in Fig. 6.18A. State with reasons the component that is faulty. How could this fault be verified?

(2) The circuit fails so that the output waveform is as shown in Fig. 6.18B. State the two components that could cause this fault. One measurement is necessary to pinpoint the fault to one of these components. State this measurement.

(3) Sketch the time-related waveforms for input and output, if the input signal were changed to pulses of 0·1 ms duration and +10 V amplitude at a frequency of 1 kHz.

(4) If the circuit fails so that no output signal is present although the input is correct, what is the first check you would make? Give your reasons.

(5) There are three components that could fail and produce the symptoms of fault (4). Which of the three is the most likely?

(6) The circuit fails so that the output is as shown in Fig. 6.18C. State the component which will cause this fault.

(7) The circuit faiis so that the output is as shown in Fig. 6.18D. Which component is faulty?

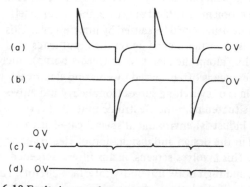

(a)
0 V
(b)
0 V
0 V
(c) -4 V
(d) 0 V

**Fig. 6.18** Faults in wave form shape

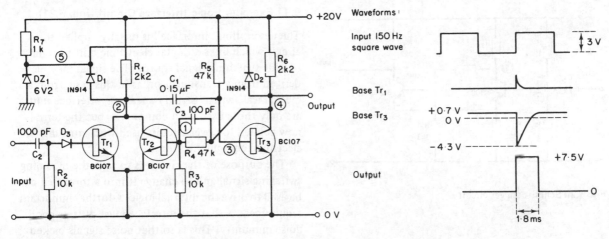

**Fig. 6.19** Monostable circuit

## 6.9 Exercise: Monostable Circuit (Fig. 6.19)

This monostable, of standard design, produces positive output pulses of about 1·8 ms duration with an amplitude of 7·5 V. Catching diodes are used ($D_1$ and $D_2$) to limit the voltage swing at both collectors. You will notice that an additional transistor ($Tr_1$) is used to buffer the input, thereby preventing any changes of state in the monostable being fed back to effect the input driving device. The waveform diagrams shown with the circuit show that a suitable test signal is a 150 Hz 3 V pk–pk square wave. This is differentiated hard by $C_2R_2$, and the positive "spike" is used to trigger the circuit into operation. The operation of the circuit has already been described in a previous section of this chapter.

### Questions

(1) Which two components determine the width of the output pulse?
(2) What is the purpose of $C_3$?
(3) In the stable state with no input applied, what voltages would you expect to measure with a 20 kΩ/V meter at all test points?
(4) How would you test that $Tr_3$ operated correctly without removing it from the circuit?
(5) In each of the following cases the monostable fails to produce an output when the correct input is applied. State which component (or components) is at fault. (+ve means "just +ve".)

(6) State the possible effect on the circuit for the following component faults:
(a) $C_2$ open circuit
(b) $Tr_1$ base emitter short circuit
(c) $Tr_3$ collector open circuit.
(7) The circuit fails so that the output pulse amplitude increases to +20 V although the pulse width remains at 1·8 ms. State, with reasons, the component fault.
(8) How could the circuit be modified to produce negative going output pulses?

|   | 1 | 2 | 3 | 4 | 5 |
|---|---|---|---|---|---|
| A | +ve | 0 | 0·7 | 0·1 | 6·2 |
| B | +ve | 7·0 | 0·7 | 0·1 | 6·2 |
| C | 0 | 7·0 | 0·7 | 0·1 | 6·2 |
| D | +ve | 0·7 | 0·7 | 0·1 | 0 |
| E | 0·7 | 0·13 | 0 | 7·0 | 6·2 |

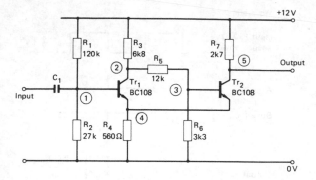

**Fig. 6.20** Schmitt trigger circuit

### 6.10 Exercise: Schmitt Trigger Circuit (Fig. 6.20)

The circuit is designed to produce square waves from
a 1 kHz 1 V pk–pk sine wave input. The potential
divider $R_1$, $R_2$ sets up a voltage on the base of $Tr_1$
that just exceeds the threshold of the circuit. Thus
in the stable state $Tr_1$ conducts and $Tr_2$ is held off.
Only when the sine wave input actually falls below
about −0·2 V will the circuit change state with $Tr_1$
OFF and $Tr_2$ ON. The circuit continuously switches
back and forth as the sine wave input changes, but
since the switching is so rapid a square wave output
is produced. The amplitude is about 5 V.

### Questions

(1) Calculate the voltages you would expect to
measure with a 20 kΩ/V meter on the 10 V range
at each of the test points. Assume no input is applied.
(2) What effect on d.c. voltages and output waveform
would result if $R_1$ went open circuit?
(3) In each of the following faults the circuit fails to
produce an output signal when the correct input is
applied. State, with reasons, the faulty component.

| | 1 | 2 | 3 | 4 | 5 |
|---|---|---|---|---|---|
| Fault | | | | | |
| A | 1·5 | 8·1 | 1·8 | 1·15 | 6·4 |
| B | 1·45 | 0·95 | 0·4 | 0·85 | 12 |
| C | 2 | 8·15 | 1·7 | 1·7 | 12 |
| D | 1·4 | 0·85 | 0·8 | 0·8 | 12 |
| E | 1·6 | 1·6 | 0·2 | 0·9 | 12 |
| F | 1·45 | 0·95 | 0 | 0·85 | 12 |

### 6.11 Exercise: Logic Interface Circuit (Fig. 6.21)

The description "interface" is usually applied to a
circuit which links a digital electronic unit to some
outside switching signal source. The signal may be
derived from a transducer or, as in this case, from a
mechanical switch. Strictly speaking, interface units
are only those used with computers, but the term is
now applied to cover circuits used with most digital
electronic equipment.

The purpose of the circuit is to take the incoming
switching signal and to change it into a form that can
be used to drive the internal logic. A further important
requirement is that the interface must possess good
noise immunity. This is so that noise signals picked
up on the input lead can be rejected. For logic
systems in very noisy electrical situations the interface unit is usually a differential receiver. The
switching signal is generated by a differential driver,
and then carried along a twisted pair of wires to the
interface receiver (Fig. 6.22). Any noise signal coupled
on to the input wires will be in common mode and
will, therefore, be rejected by the differential receiver
amplifier.

Where a single signal wire has to be used, the
circuit in this exercise does provide a high degree of
noise immunity, and also has filter circuits to remove
very high frequency noise. $L_1$, $C_1$ provide the first
low pass filter; additional filtering comes with $R_1C_2$.
The d.c. level from the switch is used to drive a
transistor inverter, and the output from this inverter
triggers a Schmitt to provide a fast logic edge to the
internal circuits. The threshold at the interface input
is arranged to be +15 V. This is the d.c. level at
which the output changes state. You can readily see
that the transistor will start to conduct when the input
is about +15 V, because then the base voltage will
just start to go positive. The switch voltage is +30 V
for logic "1" and zero volts for logic "0". A noise
pulse must be equal or greater than 15 V to cause the
transistor to change state.

Diode $D_1$ is included to protect the base emitter
junction against excessive reverse voltage. This may
be caused by a large negative noise pulse or if the
switch signal line became open circuit.

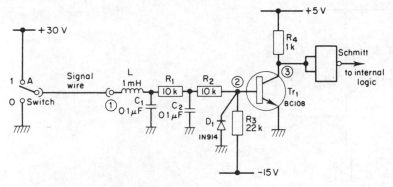

**Fig. 6.21** Logic interface unit

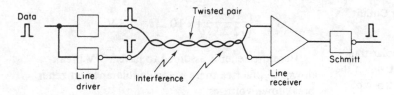

**Fig. 6.22** Typical arrangement of a differential line driver and receiver for use in very noisy electrical environments

## Questions

(1) How could you quickly check that the transistor switch is operating correctly without operating the external switch?

(2) What would be the effect on the circuit if $R_3$ became open circuit?

(3) In each of the following cases the interface fails to produce an output change of state when the external switch is operated. State which component is faulty.

|   |         | 1  | 2    | 3 |
|---|---------|----|------|---|
| A | Logic 1 | 30 | −0·6 | 5 |
|   | Logic 0 | 0  | −0·6 | 5 |
| B | Logic 1 | 30 | 0·7  | 0 |
|   | Logic 0 | 0  | −0·6 | 0 |
| C | Logic 1 | 30 | 0    | 5 |
|   | Logic 0 | 0  | 0    | 5 |
| D | Logic 1 | 30 | 5·8  | 5 |
|   | Logic 0 | 0  | −0·6 | 5 |

(4) Write down the voltages you would expect to measure using a standard multimeter if the input signal lead became open circuit.

(5) Write down the voltages you would expect to measure if the base collector junction became short circuit.

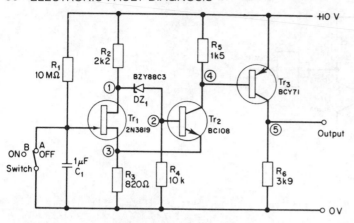

**Fig. 6.23** FET input Schmitt trigger circuit

## 6.12 Exercise: FET Input Schmitt Trigger Circuit (Fig. 6.23)

One of the major disadvantages of the bipolar transistor version of the Schmitt trigger is that, when the input transistor conducts, the input impedance of the circuit falls to a relatively low value. This is because a small base current is required for the input transistor. By using a field effect transistor a high value of input impedance can be achieved for both states of the circuit.

In the example an n-channel junction FET (type 2N3819 or equivalent) is used. With the switch closed $C_1$ is discharged and the gate potential will be zero. At the same time the source is held positive by a voltage greater than the cut-off voltage of the FET, thus ensuring that the FET is non-conducting. The circuit is wired conventionally with the source of $Tr_1$ connected to the emitter of $Tr_2$, and this transistor is forward biased by $R_2$ and $DZ_1$. Under these conditions the voltages at all the test points can readily be estimated using the following formulae and at the same time making any reasonable assumptions.

The voltage at TP2 is given by

$$V_2 \simeq \left(\frac{V_{CC} - V_Z}{R_2 + R_4}\right) R_4$$

Here we have neglected the effect of $Tr_2$ base current and also the slope resistance of the zener diode. Both these would only have a marginal effect on the calculated value, but would make the calculation much more complicated.

$$V_2 = \left(\frac{10 - 3}{12 \cdot 2 \text{ k}\Omega}\right) \times 10 \text{ k}\Omega = 5 \cdot 8 \text{ V}$$

Here the zener is assumed to have a 3 V breakdown. In practice there is a wide tolerance on zener breakdown voltages.

Voltage at TP1 (drain of $Tr_1$) is

$$V_1 = V_2 + V_{DZ} = 8 \cdot 8 \text{ V}$$

Voltage at TP3 ($Tr_1$ source and $Tr_2$ emitter) will be 0·7 V less than the base voltage of $Tr_2$.

$$V_3 = V_2 - V_{BE} = 5 \cdot 8 - 0 \cdot 7 = 5 \cdot 1 \text{ V}$$

This voltage is sufficient to ensure that the FET is fully cut off.

$Tr_2$ is conducting, so its collector voltage falls and turns on $Tr_3$. Note that $Tr_3$ is connected as a saturated switch so the voltages at TP4 and TP5 should be 9·3 V and 9·9 V respectively.

When the input switch is opened, $C_1$ can charge via $R_1$, so the gate voltage of the FET rises positive. At some point the bias voltage between gate and source has fallen to a value that allows a small drain current to start flowing. The voltage at the drain falls and so does the voltage at $Tr_2$ base. The positive feedback in the circuit ensures that $Tr_1$ rapidly turns ON and $Tr_2$ OFF. Consequently $Tr_3$ also turns OFF since it no longer has forward bias supplied from $Tr_2$ collector. The output falls to zero volts.

Because of the very high input impedance this circuit can be used to provide a long time delay between the operation of the input switch and the resulting change of state at the output. With the values of $R_1$ and $C_1$ as specified the time delay is about 5 sec. Delays of up to several minutes are possible by increasing the values of $R_1$ and $C_1$. Note that $C_1$ should be a type of capacitor that has low leakage such as a tantalum.

One of the drawbacks of this circuit is that the value of cut-off voltage for the FET has a wide tolerance and this means that the threshold level is difficult to define accurately.

The voltages at the test points were measured using a standard multimeter for both positions of the switch. In the latter case the readings were taken after a 10 sec delay.

*Normal readings*

|            | 1   | 2   | 3   | 4   | 5   |
|------------|-----|-----|-----|-----|-----|
| Switch pos. A | 8·5 | 6·1 | 5·5 | 9·2 | 9·9 |
| Switch pos. B | 3   | 1·2 | 2·5 | 10  | 0   |

## Questions

(1) If it is required to measure the gate-to-source bias voltage, what type of voltmeter should be used?

(2) The circuit fails to change state when the switch is made to B, even after several minutes. The voltages remain as for the normal readings in switch position A. Which components could cause this fault? In each case describe the type of fault and a measuring method that would confirm it.

(3) Describe the effect on the circuit operation and on the bias voltages of a base emitter short in $Tr_3$.

(4) In each of the following cases the circuit fails to change state after the operation of the switch. From the readings determine which component (or components) is faulty and the type of fault. In each case give a supporting reason.

(5) The circuit operation changes so that the delay time increases and is erratic. The readings are as given below. Which component has failed? Give reasons.

|            | 1   | 2   | 3   | 4   | 5   |
|------------|-----|-----|-----|-----|-----|
| Switch pos. A | 9·8 | 8·2 | 7·7 | 9·2 | 9·9 |
| Switch pos. B | 3   | 1·8 | 2·5 | 10  | 0   |

(6) If the switch was situated remote from the circuit, describe a simple method for checking *all* the active components without unsoldering any leads.

|         |              | 1   | 2   | 3   | 4  | 5   |
|---------|--------------|-----|-----|-----|----|-----|
| Fault X | Switch pos. A | 5   | 0   | 1·8 | 10 | 0   |
|         | Switch pos. B | 3·1 | 0   | 2·6 | 10 | 0   |
| Fault Y | Switch pos. A | 9·8 | 9·2 | 8·5 | 9·2 | 9·9 |
|         | Switch pos. B | 8·6 | 9·2 | 8·5 | 9·2 | 9·9 |
| Fault Z | Switch pos. A | 2·5 | 0·8 | 0   | 9·2 | 9·9 |
|         | Switch pos. B | 9·8 | 8·2 | 7·7 | 9·2 | 9·9 |

# 7 Thyristor and Triac Circuits

Thyristors, formerly called silicon controlled rectifiers (SCRs), and triacs are semiconductor devices that act as high-speed power switches. Devices are available that can operate at potentials of several hundred volts and which will carry currents of up to hundreds of amperes. These solid state units are now being increasingly used to replace conventional mechanical switches and relays since they offer faster switching and greater reliability. This is particularly the case in continuous a.c. power control systems such as lamp dimmers, heater control, motor speed control, etc. This is a growing and important branch of electronics, so the understanding of the operation of the devices, their use, and fault diagnosis is also important. We shall start by outlining the basic operation.

## 7.1. Principle of Operation of the Thyristor

The thyristor has a structure (Fig. 7.1) that consists of a four-layer p-n/p-n silicon sandwich. The symbol is that of a rectifier with an additional terminal called the GATE. It is this gate which enables the action of the rectifier to be controlled. The device can be made to act as an open circuit (forward blocking) or it can be triggered into a low resistance forward conducting state by applying a short pulse of relatively low power to the gate terminal.

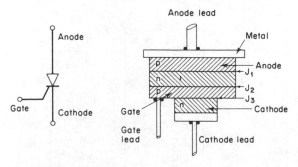

**Fig. 7.1** Construction of a thyristor

One of the aids in understanding the operation is the two transistor equivalent circuit (Fig. 7.2). By dividing the thyristor diagonally it can be seen that a p-n-p transistor structure exists between anode and gate, and an n-p-n transistor in the gate cathode region.

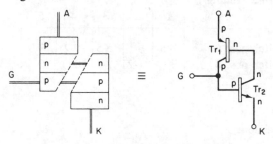

**Fig. 7.2** Two transistor equivalent circuit of a thyristor

The operation can be divided under the following sub-headings:

(*a*) *Reverse bias,* anode negative with respect to cathode. The thyristor is in reverse blocking state and only a low-value leakage current flows. Under these conditions both junctions $J_1$ and $J_3$ are reverse biased.

(*b*) *Forward bias,* anode positive with respect to cathode but *no gate signal.* Thyristor is said to be forward blocking since it acts as a high resistance. Only a small leakage current flows. It can be seen that although $J_1$ and $J_3$ are forward biased, the centre junction $J_2$ is reversed. By referring to the equivalent circuit one can explain forward blocking by the fact that, since the gate has no signal applied to it, $Tr_2$ is cut off. Only a small leakage current can flow.

(*c*) *Forward bias with gate signal applied.* If a pulse of forward bias is applied between gate and cathode while the anode is positive with respect to the cathode then the thyristor will be forced into conduction. The switch-on time is rapid (microsecs) and a large current can be passed by the device, limited only by the external resistance. The anode to cathode voltage falls to a low value, typically 1 V. This action can be explained using the equivalent circuit by noting that a pulse of forward bias turns on $Tr_2$. This transistor starts to conduct and therefore switches on $Tr_1$. The two transistor circuit has a positive feedback loop, since each has its collector wired to the other base. Therefore the two transistors rapidly switch on, and will remain on even when the gate signal is removed. The device can only be turned off by reducing the anode current below a value known as the "holding current".

Since the two transistors are connected as a positive feedback pair there is a large current gain between the gate ($Tr_2$ base) and the anode ($Tr_1$ collector and $Tr_2$ emitter current). The voltage gain is also high since only about 1 V is required between gate and cathode to trigger the thyristor on. The amount of gate power required to turn on a typical thyristor is thus relatively small. In fact, milliwatts of gate power can be used to control hundreds of watts of load power in the anode circuit.

Mention has been made of the important fact that once triggered the thyristor remains conducting even when the gate signal is removed. It can only be switched off by reducing the current flowing through it to a value less than the holding current. This is the specified minimum current that will ensure continued conduction, and it is usually a few per-cent of the maximum forward current. In a.c. circuits the thyristor turns off every half cycle when the supply voltage passes through zero and goes negative; this gives automatic turn off. In d.c. circuits special techniques are used to reduce the anode current to zero to switch off the device.

There are two other conditions, apart from gate signal, that will switch a forward blocking thyristor into conduction:

(a) Exceeding the forward breakover voltage.
(b) By applying a fast rising voltage wave form between anode and cathode, typically greater than 50 V per microsec. This "rate-effect" is explained by the fact that internal capacitance ($J_2$ junction) can feed a part of a sharply rising anode voltage through to the gate. This turns the thyristor on.

Both these effects are undesirable.

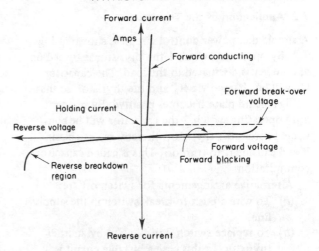

Fig. 7.3 Typical static characteristics of a thyristor

Another point worth noting concerning circuits using high-speed power switches such as thyristors is that, since the switch-on is so rapid, switching "spikes" can easily be generated. Therefore most circuits, especially a.c. controllers, include RF filters to limit the amount of RF interference on the a.c. power line.

## 7.2. Application of the Thyristor

A simple d.c. power control circuit is shown in Fig. 7.4. By operating switch 1 the thyristor is turned on and power is dissipated in the load. The capacitor $C_1$ will be charged via $R_3$ and the thyristor, so that its right-hand plate becomes positive. By momentarily operating switch 2 the capacitor will be connected across the thyristor, thereby reverse biasing it and causing it to turn off. This is called capacitor commutation.

Alternative arrangements for switch off are:

(*a*)  To wire a push-to-break switch in the supply line.

(*b*)  To replace switch 2 in Fig. 7.4 by another thyristor. In this case a bistable circuit is formed. Such a circuit (Fig. 7.5) can be used as a high-speed trip.

One of the important uses of thyristors is in the smooth control of a.c. power (Fig. 7.6). The average power in the load can be varied by adjusting the time position of the gate triggering pulse relative to the supply waveform. Conduction angles from $10°$ to $170°$ are possible. As the supply voltage goes positive $C_1$ is charged via $R_1$ and the variable resistor $RV_2$. When the potential across $C_1$ exceeds the breakover voltage of the trigger diode (diac) then the diac conducts and supplies a gate pulse to the thyristor. The thyristor switches on, and nearly the whole of the supply voltage is applied across the load. By adjusting $RV_2$ the charging time of $C_1$ can be varied and this in turn varies the conduction angle. The diode $D_1$ is included to prevent reverse bias being applied to the thyristor gate, and at the same time it ensures stable triggering by discharging $C_1$ on each negative half cycle.

Because of its low efficiency this simple half wave circuit is not suitable for most application. Full wave controllers using a bridge rectifier (Fig. 7.7) with two thyristors, or better still one triac, being preferred.

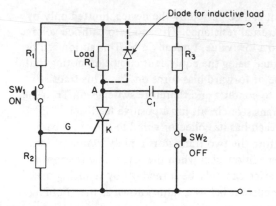

Fig. 7.4 A simple d.c. power control circuit using a thyristor

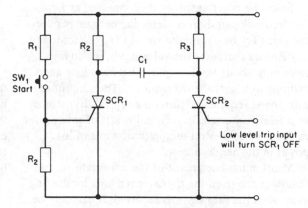

Fig. 7.5 Using a slave thyristor as the turn-off device in a d.c. circuit

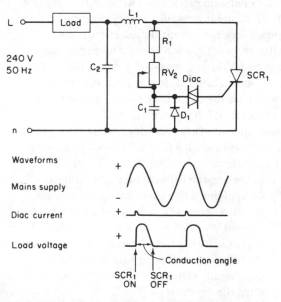

Fig. 7.6 Half wave a.c. power control using a thyristor

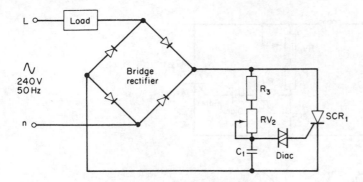

**Fig. 7.7** Full wave a.c. controller using a thyristor

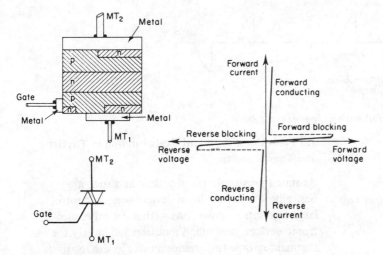

**Fig. 7.8** Basic construction and static characteristics of a triac

## 7.3 Basic Operating Principle of the Triac

The construction of this device is shown in Fig. 7.8 where it can be seen that it may be regarded as two thyristors connected in inverse parallel, with a common gate terminal. The operation is similar to that of the thyristor.

The triac will therefore pass or block current in both directions and it can be triggered into conduction in either direction by positive or negative gate signals. The characteristics are also shown in Fig. 7.8.

The four triggering modes are signified as follows:

| | | |
|---|---|---|
| $I^+$ mode = MT2 current +ve | Gate current +ve |
| $I^-$ mode = | +ve | −ve |
| $III^+$ mode = | −ve | +ve |
| $III^-$ mode = | −ve | −ve |

The highest sensitivity is achieved in the $I^+$ and $III^-$ modes, these being about equal and twice that of the other two modes.

As for thyristors only a small amount of gate power is required to turn the device on which will then control large amounts of power in its MT2 circuit.

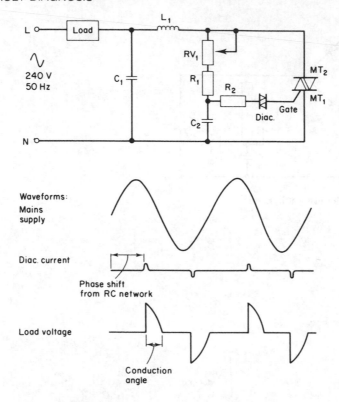

Fig. 7.9 Typical triac full wave a.c. power controller

## 7.4 Applications of the Triac

Triacs are mostly used in full wave a.c. control circuits since that is the area in which they offer two distinct advantages over two thyristors:

(a) Simpler heat sink design.

(b) Relatively economical trigger circuits.

A typical triac a.c. controller with diac triggering is shown in Fig. 7.9. The conduction angle is controlled by the potentiometer, and the gate receives positive current on the positive half cycle of the supply and negative current on the negative half of the supply ($I^+$ and $III^-$ modes). The waveforms show that full wave control is achieved.

## 7.5 Fault Conditions and Fault Finding in Thyristor and Triac Circuits

As stated previously the thyristor and triac are basically four-layer silicon devices used to control large amounts of power. As with most other electronic devices, controlled rectifiers fail largely for thermal reasons. High temperatures in the relatively small volume of the device, or a high rate of temperature cycling, causes slow deterioration which ultimately leads to failure.

Controlled rectifiers can also be destroyed like fuses if they are subjected to overload current surges. Naturally it is the designer's job to ensure that the device is mounted on to an efficient heat sink and that overload current surges do not occur.

The faults that can occur are low voltage forward breakover, loss of gate control, open or short circuit anode to cathode, or open or short circuit gate to cathode.

It is also possible for thyristors and triacs to fail by what is called $di/dt$ failure. This is explained by considering the instant at which the device is triggered. Gate current is concentrated in a very small area of the gate region. As a result the initial flow of anode current is constrained into a small area, and if this anode current has a rate of change ($di/dt$) that exceeds a critical value, then a large amount of heat will be generated in a small area (hot spot) and the device will fail. In most circuits the load inductance limits the rate of change of current.

Mention has already been made of the $dv/dt$ effect, whereby a thyristor can be triggered on by a sharply rising anode voltage. You will find some circuits with an RC network wired in parallel with the thyristor or triac to eliminate this effect.

The effects of various faults can be more easily understood by using the two transistor equivalent circuit.

In most cases the symptoms and the voltages in the circuit will indicate the fault. But always remember that the device may be used in a d.c. or an a.c. circuit. For example a thyristor may be used to control the power in a load connected to the 240 V 50 Hz mains with a d.c. signal applied to its gate. Check with the circuit diagram before making any measurements.

With some circuits it is possible to test the thyristor or triac while it remains in circuit. For example it is quite reasonable to inject a suitable trigger signal into the gate to determine if it is the thyristor or the gate signal source that has failed in a circuit. Also, with the power supply OFF, measurements with an ohmmeter can be made to test for short circuit anode to cathode or for open or short gate to cathode. The gate cathode junction has the same characteristics as a diode. A resistance of about 500 Ω should be indicated in the forward direction (gate +ve w.r.t. cathode) and a high resistance (greater than 100 kΩ) in the reverse direction. Remember that, with the device in circuit, other components in parallel with the gate will affect the readings.

A simple circuit can be used to check the operation of a thyristor (or triac). The test circuit of Fig. 7.10 can indicate gate operation, forward leakage current, forward voltage drop, and minimum holding current. With $R_2$ set to minimum $S_1$ is closed. The indication of meter 1 should be very low (typically less than 50 $\mu$A) and meter 2 should indicate nearly 12 V.

TABLE 7.1 Typical Faults on Thyristor Circuits

| FAULT | RESULT AND SYMPTOMS |
|---|---|
| Gate to cathode open circuit | Thyristor OFF and cannot be triggered into conduction. Measured gate signal high. |
| Gate to cathode short circuit | Thyristor OFF and cannot be triggered into conduction. Measured gate signal is zero. |
| Anode to cathode short circuit | Thyristor conducting in both forward and reverse directions. Measured volt drop between anode and cathode is zero. |
| Anode or cathode open circuit | Thyristor OFF. |

Note that if the actual leakage current of the thyristor in the forward blocking mode is required then meter 2 should be disconnected.

Next depress switch 2 for a short time; this should trigger the thyristor into conduction. Meter 1 should indicate approximately 100 mA and meter 2 approximately 1 V. The latter is the forward voltage drop.

To obtain a value for the minimum holding current gradually increase $R_2$ until the thyristor turns off. The current indicated just before turn off is the minimum holding current.

The circuit should be modified for tests on high power devices by using lower value resistors.

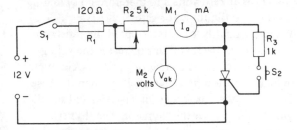

**Fig. 7.10** A d.c. test circuit for medium power thyristor

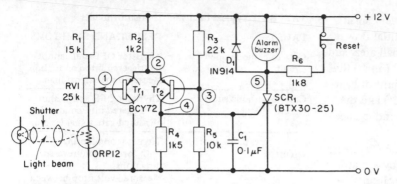

**Fig. 7.11** Alarm unit

### 7.6 Exercise: Alarm Unit (Fig. 7.11)

This circuit shows the use of a thyristor in a d.c. mode. If the light beam is interrupted, the thyristor is switched on to operate the alarm buzzer. The buzzer would normally turn the thyristor off since it switches open every time it operates, but a holding current path is provided via $R_6$. The diode $D_1$ is included to protect the thyristor from the large back e.m.f. spikes generated as the buzzer switches.

Tr$_1$ and Tr$_2$ form a differential amplifier which amplifies the difference signal between Tr$_1$ and Tr$_2$ bases. Tr$_2$ base is held at a constant voltage by the potential divider $R_3$ and $R_5$, while the input to Tr$_1$ base depends upon the resistance of the photocell and the setting of RV$_1$. When the light beam is fully on the photocell, this has a low resistance so that the voltage on Tr$_1$ base is lower than that on Tr$_2$ base. Tr$_1$ conducts and Tr$_2$ is held off. The collector voltage of Tr$_2$ is at zero volts and therefore no switching signal is fed to the thyristor gate.

When the light beam is interrupted the resistance of the photocell rises, and this increases the voltage on Tr$_1$ base. Tr$_1$ stops conducting and Tr$_2$ conducts since its base is now at a lower voltage than the base of Tr$_1$. The collector voltage of Tr$_2$ rises and supplies a signal to the gate of the thyristor. The thyristor turns on and operates the buzzer indicating that the light beam has been momentarily broken. The circuit can be reset by operating switch 1.

### Questions

(1) State the voltages you would expect to measure with a 20 k$\Omega$/V meter with the light beam *on* and *off* at the various test points.

(2) What would be the symptoms for an anode to cathode short circuit of the thyristor?

(3) In the following faults the voltages were measured using a standard meter. State with a supporting reason the faulty component or components.

(A) The light beam is cut off but the alarm fails to operate.

| TP | 1 | 2 | 3 | 4 | 5 |
|----|---|---|---|---|---|
| MR | 3.9 | 4.4 | 3.7 | 4.3 | 12 |

(B) The light beam is cut off but the alarm fails to operate.

| TP | 1 | 2 | 3 | 4 | 5 |
|----|---|---|---|---|---|
| MR | 0.7 | 1.3 | 3.7 | 0 | 12 |

(C) The alarm operates continuously and cannot be reset.

| TP | 1 | 2 | 3 | 4 | 5 |
|----|---|---|---|---|---|
| MR | 3.1 | 3.7 | 3.7 | 0 | 0 |

(D) The alarm fails to operate when the light beam is cut off.

| TP | 1 | 2 | 3 | 4 | 5 |
|----|---|---|---|---|---|
| MR | 4.2 | 4.8 | 11.5 | 0 | 12 |

(E) Interrupting the light beam fails to trigger the alarm.

| TP | 1 | 2 | 3 | 4 | 5 |
|----|---|---|---|---|---|
| MR | 0 | 0.7 | 3.7 | 0 | 12 |

(F) The alarm operates continuously and cannot be reset.

| TP | 1 | 2 | 3 | 4 | 5 |
|----|---|---|---|---|---|
| MR | 2.5 | 1.8 | 0.75 | 0.75 | 0.95 |

(4) If the alarm operates once, then fails to hold, when the light beam is momentarily cut, which components could be considered faulty? What simple method could be used to verify, and to locate, the fault to one component only?

(5) Write down the series of simple, rapid tests that could be made to verify the operation of various components in the circuit when no multimeter is available.

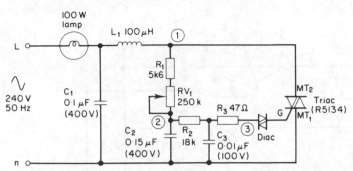

**Fig. 7.12** Lamp dimmer

## 7.7 Exercise: Lamp Dimming Circuit (Fig. 7.12)

As explained previously in section 7.4 the triac is an excellent device for full wave a.c. power control. This circuit uses the phase triggering technique, in which $RV_1$ and $C_2$ provide a variable potential divider and variable phase shift network. This feeds a phase shifted signal to a "slave" network $R_2C_3$. When the voltage across $C_3$ exceeds about 35 V the diac triggers to partially discharge $C_3$ into the triac gate. The triac then conducts and power is applied to the lamp. The purpose of the "slave" network is to prevent any large change of voltage occuring across $C_2$ when the diac triggers.

The conduction of the triac can be controlled over a wide angle by adjusting $RV_1$. For example, with $RV_1$ set to minimum very little phase shift or potential divider action takes place and the voltage across $C_2$ (and $C_3$) almost follows that of the a.c. supply rail. The diac therefore passes a trigger pulse to the triac gate shortly after the start of each half cycle of the supply. You can see that almost full power is applied to the lamp.

When $RV_1$ is adjusted to near maximum value the phase shift in the RC network approaches $90°$. From basic a.c. theory the phase angle for an RC network is

$$\tan \theta = \omega CR$$

So with $R = 250$ k$\Omega$, then $\tan \theta = 2\pi50 \times 0.15 \times$

$10^{-6} \times 250 \times 10^{-3}$ and $\theta = 85°$.

The attenuation in the network with $R = 250$ k$\Omega$ can be calculated from

$$v_c = \frac{V_s}{Z} X_c$$

where impedance $Z = \sqrt{(R^2 + X_c^2)}$ and $X_c = \frac{1}{2\pi fC}$.

You can readily calculate that when $RV_1$ is nearing its maximum value that the voltage across $C_3$ just reaches the voltage required to trigger the diac. This coupled with the near $90°$ phase shift means that the triac firing signal is delayed by nearly $170°$, and the lamp therefore has power applied to it for only $10°$ of each half cycle.

## Questions

(1) What is the purpose of $L_1$ and $C_1$?

(2) If the r.m.s. current rating of the triac used is 6 A, what is the maximum load power that can be controlled?

(3) What would be the symptoms if $RV_1$ wiper became open circuit?

(4) A fault exists such that the lamp is at full brightness and cannot be controlled by adjusting $RV_1$. The a.c. Voltage between TP1 and the neutral line measures zero. State the component failure.

(5) The lamp fails to light. The a.c. voltages, with $RV_1$ set to minimum, are as follows:

| TP | 1 | 2 | 3 |
|---|---|---|---|
| A.C. r.m.s. voltage | 235 | 227 | 0 |

State the component failure. How could this fault be quickly verified?

(6) The unit fails so that the control over the lamp's brightness becomes limited. The lamp burns at high brightness with $RV_1$ at minimum but only reduces slightly in intensity with $RV_1$ at maximum. Which component has failed? How could this be quickly checked?

(7) What symptoms would indicate a gate to cathode open circuit? How would they differ from those of a gate to cathode short circuit?

(8) The lamp fails to light at all. The voltage at TP1 is zero. Which component has failed?

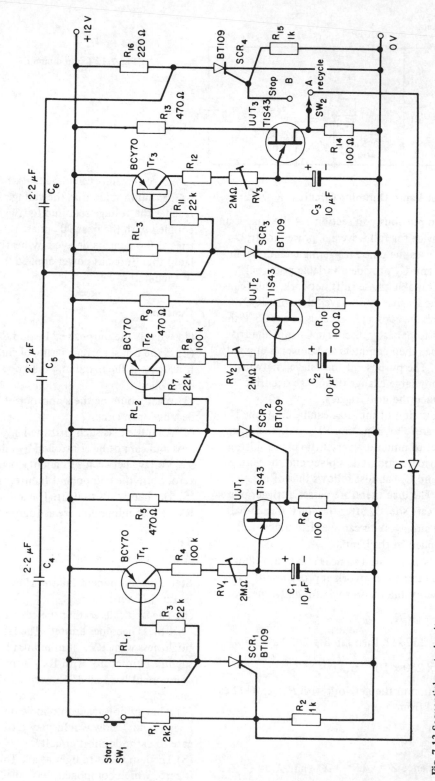

**Fig. 7.13** Sequential control unit
NB $C_4$, $C_5$ and $C_6$ must be non-polarized types

### 7.8 Exercise: Sequential Control Unit (Fig. 7.13)

In many process control situations a circuit is required that can switch on various loads in a defined sequence, each load being on for a controlled period of time. For example a sequence of control may be required as follows:

| Start | Load 1 | Operate belt mechanism for 5 sec to move work into position. |
|---|---|---|
| | Load 2 | Spray work for 2 sec. |
| | Load 3 | Heat for 10 sec. Then stop. |

In the circuit being considered the thyristors are used to switch power to the various loads and unijunctions are used to provide the time delay. The circuit consists of three identical stages.

When power is applied none of the thyristors conduct since they do not receive gate signals. By pressing the start switch a gate signal is provided for $SCR_1$ which switches into forward conduction and connects power to load $L_1$. Since the anode voltage of $SCR_1$ falls to about +1 V, $Tr_1$, which is a pnp transistor, is forward biased via $R_3$. This transistor conducts and $C_1$ is charged via $R_4$ and $RV_1$ towards the positive line. When the voltage across $C_1$ equals the emitter trigger voltage of $UJT_1$, the unijunction conducts and a positive pulse is generated across $R_6$ to fire thyristor $SCR_2$. The anode voltage of $SCR_2$ falls and this feeds a negative edge via $C_4$ to the anode of $SCR_1$ to turn off $SCR_1$. Power is applied to load $L_2$ for a time determined by $RV_2$ and $C_2$. This part of the circuit is identical to that already described so that when $UJT_2$ triggers, $SCR_3$ conducts and this switches off $SCR_2$.

At the end of the sequence when $UJT_3$ triggers on, the circuit can be made to automatically recycle by placing $SW_2$ to position A. Then the pulse from $UJT_3$ is fed to the gate of $SCR_1$. With the switch $SW_2$ in position B, $SCR_4$ is made to conduct and this switches off $SCR_3$ to end the sequence. The cycle can be restarted by pressing the start button.

### Questions

(1) What type of instrument should be used to measure the voltage across $C_1$, $C_2$ or $C_3$?

(2) If the trigger voltage of the unijunction is assumed to be approximately 7 V calculate the minimum and maximum operate time of a load.

(3) The unit fails so that it will neither recycle, nor switch off power to load $L_3$ at the end of the sequence. Pressing the start button will however initiate operation. State which portion of the circuit is at fault and list the likely component failures.

(4) Suggest a circuit modification to ensure that pressing the start button switches $SCR_1$ on only when thyristors $SCR_2$ and $SCR_3$ are OFF.

(5) The unit fails so that power is applied to load 2 and load 3 as soon as load 1 is off. Loads 2 and 3 also switch off at the same instant. State, with reasons, the component fault and the type of failure.

(6) The unit fails so that it will not recycle. Operating the start button will initiate the sequence and the unit stops correctly with $SW_2$ in position B. Which component has failed?

(7) A fault exists so that load 3 is always energized when power is applied. State the probable component failures and show how one voltage measurement could be used to locate the fault.

(8) State fully the symptoms for the following faults:

   (a) $Tr_2$ base emitter open circuit

   (b) $C_2$ open circuit

   (c) $C_6$ open circuit

   (d) $SCR_1$ gate to cathode short circuit

   (e) $SCR_3$ anode to cathode open circuit.

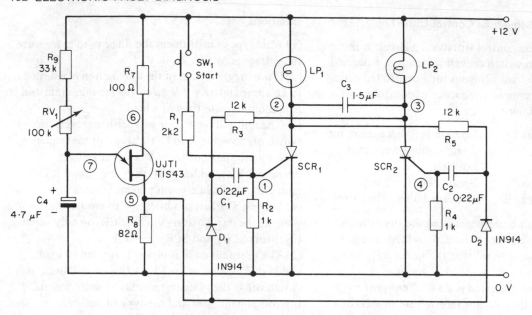

**Fig. 7.14** Lamp flasher unit

### 7.9 Exercise: Lamp Flasher Unit (Fig. 7.14)

This unit comprises a low-speed unijunction oscillator which is used to drive a thyristor bistable circuit. Each pulse from the oscillator is steered via a diode circuit to the gate of the thyristor that is off.

Initially when power is applied to the circuit both thyristors are off and although positive pulses are generated from the unijunction neither thyristor is triggered since diodes $D_1$ and $D_2$ are reverse biased. Depressing the start switch causes $SCR_1$ to conduct and lamp $LP_1$ lights. The anode of $SCR_1$ is then at approximately +1 V while that of $SCR_2$ is at +12 V. The next positive pulse from $B_1$ of the unijunction is then sufficient to forward bias $D_2$ but not $D_1$, and a trigger pulse is supplied to the gate of $SCR_2$. This thyristor switches into a forward conducting state to light lamp $LP_2$. At the same time a negative step is transmitted via the commutating capacitor $C_3$ to reverse bias and turn off $SCR_1$. The next pulse from the oscillator will be steered via $D_1$ to trigger thyristor $SCR_1$. This lights $LP_1$ and at the same time turns off $SCR_2$. The lamps therefore flash on and off alternately at a frequency determined by the unijunction pulse generator.

Fault finding in this type of circuit is usually relatively easy since a visual indication of the state of the circuit is given by the lamps. The lamps themselves are the components with the highest failure rate, so the service engineer would check these first. Assuming the unit has failed with both lamps out, a first check would be to short the anode to cathode of each thyristor in turn to test the lamps. A later question concerns the next set of checks that would be made.

#### Questions

In all cases you can assume that both lamps are functional.

(1) Calculate the approximate maximum and minimum frequency of the unijunction oscillator.

(2) A fault exists so that both lamps are lit and remain on as soon as the start button is pressed. The voltages measured with a 20 k$\Omega$/V multimeter are as follows:

| TP | 1 | 2 | 3 | 4 |
|----|-----|------|------|---|
| MR | 0·7 | 0·85 | 0·85 | 0 |

(3) The unit fails with both lamps unlit. List in a logical order the tests that should be made to locate the fault.

(4) State, with reasons, the symptoms for the following component failures.

(a) $D_1$ open circuit.

(b) $SCR_2$ gate to cathode short circuit.

(c) $C_3$ open circuit.

(d) $SCR_1$ anode to cathode open circuit.

(5) For the following fault conditions state, with a supporting reason, the component or components that are faulty. (Assume that the start button has been pressed.)

(6) Suggest a circuit modification so that the unit would self-start from a power line switch.

(7) If it was found that the failure rate of the lamps was too high for a particular application, are any modifications of a very simple nature possible to improve the situation?

| Fault | 1 | 2 | 3 | 4 | 5 | 6 | 7 |
|---|---|---|---|---|---|---|---|
| A | 0·7 | 0·85 | 12 | 0 | 0·1 | 11·8 | 0 |
| B | 0 | 12 | 0·85 | 0·7 | Varying | Varying | Varying |
| C | 0 | 12 | 12 | 0 | Varying | Varying | Varying |
| D | 0 | 12 | 0·85 | 0·7 | 5·2 | 5·2 | 5·8 |
| E | 0·7 | 0·85 | 12 | 0 | Varying | Varying | Varying |
| F | 0·7 | 0·85 | 12 | 0 | 0 | 0·1 | 0·85 |

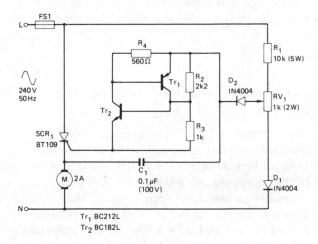

**Fig. 7.15** Motor speed control unit

### 7.10 Exercise: Motor Speed Control Circuit (Fig. 7.15)

The motor that is commonly used in appliances such as electric drills, sanders, food mixers, etc. is the series wound "universal" type electric motor. This consists of a field winding and an armature wired in series between the motor terminals. When a voltage (a.c. or d.c.) is applied across the terminals a current flows through the field winding and the armature. Opposing magnetic fields are set up between the field and the armature, and this forces the armature to rotate. As the armature rotates it generates a voltage in opposition to the voltage across the motor, and this back-e.m.f. has a magnitude that is proportional to the motors speed. The faster the armature rotates the greater the back-e.m.f. and the smaller the required motor current. This is because the current taken by the motor is proportional to the difference between the applied voltage and the back-e.m.f. When the motor is first started a large current is taken since the back-e.m.f. is zero. This means that a high torque is developed and the motor rapidly increases its speed. As it does so the back-e.m.f. increases, the current falls, and the available torque is reduced. Now if a load is applied to the motor its speed falls initially and this lowers the back-e.m.f. Therefore the current increases and this tends to restore the motor to its original speed. This type of motor then has inherent self-regulating speed properties.

In this circuit a thyristor is used to control the operating speed of the motor by switching pulses of current on positive half cycles of the mains supply. For high speed the thyristor will be triggered on very early in each positive half cycle, and by increasing the phase shift of the gate trigger signal the thyristor will be triggered later and this reduces the motor's speed.

Each positive half cycle of the supply causes a current to flow through the potential divider network $R_1$, $RV_1$, and $D_1$. An attenuated positive half cycle of the supply therefore appears on the wiper of $RV_1$ and this positive voltage charges $C_1$ via $D_2$. In fact

$C_1$ stores a charge that is proportional to the difference between $RV_1$ wiper and the voltage on the cathode of the thyristor. The latter voltage is of course the speed-dependent back e.m.f. of the motor. When the voltage across $C_1$ exceeds about 3 V, $Tr_1$ and $Tr_2$, which are wired as a regenerative switch, are triggered into conduction and supply a pulse of current to the thyristor gate. The thyristor conducts to supply power to the motor. On the negative half cycle of the mains the thyristor naturally turns off. One of the features of the universal-type motor is that by operating from a half wave rectified supply only about 20% of the power from an equivalent full wave system is lost. This fact enables quite efficient and relatively cheap half wave control systems to be used.

The circuit achieves low speed operation when the wiper of $RV_1$ is moved towards the anode of $D_1$. This means that a lower fraction of the positive half cycle is applied across $C_1$, and since the other end of $C_1$ is connected to the motor's back-e.m.f. the transistor trigger switches at a later stage in the cycle. This reduces the power switched to the motor by the thyristor and therefore reduces the motor's speed.

At very low speed, "skip-cycling" takes place, when the thyristor is used to deliver power in a portion of one half cycle out of, say, five. This occurs because the gate signal cannot, in this circuit, be later than 90° in each positive half cycle. So when the thyristor fires, it supplies at least a quarter cycle of power to the motor. This causes the motor to accelerate and to increase the back-e.m.f. Consequently the voltage across $C_1$ on the next half cycle will be too small to trigger $Tr_1$ and $Tr_2$, and the thyristor will remain off. The thyristor will only conduct again when the speed and back-e.m.f. have fallen to the original value and this may take several cycles. At first sight this would seem to be a disadvantage, but when the motor's speed is averaged out over a period of say a quarter of a second one can see that a fairly constant low speed is maintained.

Regulation at high or low speeds is achieved since any sudden change of load must produce a drop in motor speed with a consequent fall in the back-e.m.f. This means that on the next positive half cycle the thyristor receives a gate pulse earlier in the cycle and this switches on the thyristor which delivers more power to the motor than previously.

## Questions

(1) Explain briefly how the transistor trigger circuit of $Tr_1$ and $Tr_2$ works. (Hint — study the equivalent circuit of the thyristor.)
(2) The control system fails so that the motor operates at maximum speed and $RV_1$ has no control. State, with a supporting reason, the component failure.
(3) The control system fails so that the motor fails to run at all. List, with a supporting reason, the possible component failures that could cause this fault? How would you test the circuit to locate the fault to one component?
(4) State the symptoms for the following component failures:
    (a) $D_1$ open circuit.
    (b) Thyristor gate to cathode short circuit.
(5) A modification is included to reduce the effect of "skip-cycling" and judder for a particular motor by connecting a 2 $\mu$F capacitor across $RV_1$. Explain briefly how this modification affects the operation of the circuits.

### 7.11 Exercise: Isolated mains switching circuit (Fig. 7.16)

Microcomputers and digital circuits are often interfaced to loads connected in the a.c. supply. Isolation between the sensitive low voltage circuits of the microcomputer and the high voltage a.c. mains is then essential. Apart from the safety aspects a fault in the switch must not be allowed to feed the a.c. mains back into the microcomputer port. One standard method of isolation is to use one of the wide range of opto-couplers. In these units the signal is transmitted across an insulating gap by a light pulse and the input and output circuits are then completely isolated from each other. This circuit uses the OPI3041 in which the LED input is optically coupled to a light sensitive triac. The gate trigger circuit of this internal triac is controlled using a zero-crossing switch (ZCS) to ensure that the switch only operates when the a.c. mains is passing through zero volts. This technique prevents switching spikes and radio frequency interference being generated by the main triac.

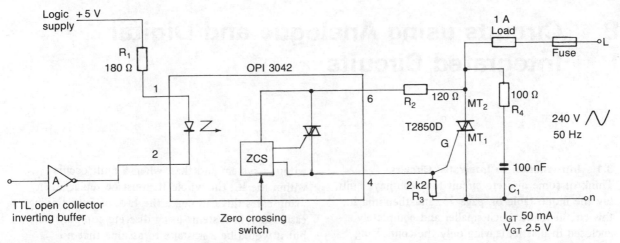

**Fig. 7.16**

This would occur if the main triac was allowed to switch when the mains was at its peak value.

The circuit operation is straightforward. When the digital input is logic 1 the output of the TTL gate A goes low and current passes through the LED via $R_1$. The infra-red light beam from the LED crosses the gap and causes the light sensitive triac to conduct at the next zero-crossing point in the mains cycle. This triac in turn passes a gate trigger pulse to the main triac (T2850D) which latches on to switch power to the load for one complete half cycle. The load continues to be connected for half cycles of mains power for as long as the logic 1 level is applied at the input. Each time the mains supply passes through zero the internal triac in the opto-isolator fires and switches the main triac on. When the logic input is 0 the LED is off and neither of the two triac conduct.

The circuit formed by $R_4$ and $C_1$, in parallel with the main triac, is called a "snubber". It acts as a filter to prevent spikes on the mains supply causing the triacs to falsely switch on.

The use of the zero-voltage switch means that power can only be applied for complete half cycles so control of power in the load is referred to a "*burst cycle*" control. For example the load could be switched on for one cycle out of every 100 cycles for 1% power, 20 cycles out of every 100 cycles for 20% power and so on. Such control can only be used with loads, such as heaters, that respond slowly in comparison with the time taken for 100 mains cycles.

**Questions**

(1) Assuming that the LED in the opto-coupler has $V_F = 2 \cdot 2$ V calculate the forward current $I_F$ of the LED when the logic input is high.

(2) A repetitive test waveform from a micro-computer is used to switch the load on for 5 mains cycles and off for 20 cycles. Determine

    (*a*)  The time the digital signal is high (1)

    (*b*)  The time the digital signal is low (0)

    (*c*)  The frequency of the signal

    (*d*)  The mark to space ratio of the signal.

(3) The unit fails to operate when a logic 1 is applied. Explain the sequence of tests you would make to locate the fault.

(4) Describe the differences in fault symptoms for the following

    (*a*)  $R_1$ open circuit

    (*b*)  the LED open circuit

    (*c*)  Gate A output stuck at 0.

(5) The unit fails with the load on continuously. The d.c. voltage reading at gate A output is $4 \cdot 9$ V. State the portion of the circuit that is at fault, and the likely fault condition. Describe any tests you could make to narrow down the fault to one component.

# 8 Circuits using Analogue and Digital Integrated Circuits

## 8.1 Introduction to Integrated Circuits

Think of some discrete circuit that you have built, say the monostable on page 75, and then imagine this circuit made much smaller and completely enclosed in plastic, leaving only the connecting pins showing. What we have then is an integrated circuit, i.e. an encapsulated unit containing all the necessary diodes, transistors and other components for a particular function. The first integrated circuits were in fact made in this way; but the mass-produced ICs of today are nearly all silicon monolithic types. In a monolithic IC, all the elements are diffused and interconnected in one piece of silicon. This small piece of silicon is referred to as the *chip*. The word monolithic comes from the Greek "mono", meaning single, and "lithos", meaning stone (in this case the stone is silicon). Other types of IC include the film circuits (both thick and thin films). In these, conducting and resistive tracks are formed on the surface of an insulating inert substrate. To complete the circuit, tiny active components such as diodes and transistors are bonded in position and the unit is then encapsulated. In general film ICs are used where the ratio of passive to active devices is fairly high.

The advantages of ICs over discrete circuits are mostly due to the fact that many more active elements can be fitted into a small space. This high component packing density gives low cost, higher reliability, and the possibility of producing circuits that could hardly be justified using discretes. From small-scale integration has grown the medium-scale integrated circuits (MSI); then the large-scale integrated circuits (LSI); leading up to very-large-scale integration (VLSI) used on the larger microprocessor and memory ICs. A VLSI circuit may contain more than 100 000 transistors.

It follows from the preceding paragraph that the IC must be considered as a functioning block within a system and that, when a fault occurs within the IC, the whole IC must be replaced. In many cases this can make the task of fault finding easier than in systems using discrete components; but it would be a mistake to assume that no knowledge of the internal operation of the IC is required. Having a good understanding of the function of the ICs in a system and the way in which they work is essential. It enables the fault-finder to sort out faults that are definitely not the IC and therefore prevents the unsoldering and replacing of a perfectly good IC when the fault is either an external component or a bad connection.

Integrated circuits are usually sub-divided into two main areas:

**Analogue ICs:** those used to amplify, process or operate on input signals that can vary anywhere between defined limits.

**Digital ICs:** the type used in logic and computer systems where the inputs are usually either high (logic 1) or low (logic 0) but not any value between these states.

A table can be compiled as in Table 8.1 to show the distinction between IC types, but it should be noted that not all types are included and also that there may be some overlap.

TABLE 8.1

| Linear ICs | Digital ICs | |
|---|---|---|
| Operational amplifiers (typical 741) | DTL | Logic gates AND/OR, NAND/NOR. |
| Audio amplifiers (typical LM380) | TTL | Schmitts. Monostables. |
| RF amplifiers | ECL | Bistables. Counters. |
| Wideband amplifiers | | Shift registers. |
| Power amplifiers (with heat-sink attached) | MOS | Memories. Clocks. |
| Voltage regulators | CMOS | Microprocessors. |
| Demodulators | | |

## 8.2   Analogue ICs

A true analogue signal is one that, at any instant in time, can be any value within a defined range. For example the output of a microphone is an electrical signal that is analogous to the input sound wave; in other words the microphone output signal varies in time in a similar fashion to the sound wave. A circuit used to amplify this small signal must be an analogue type in order to preserve the waveshape. Analogue ICs are also sometimes referred to as LINEARS since the response required from the circuit has to be reasonably linear in order to avoid distorting the signal as it is being amplified.

By far the most commonly used analogue ICs are OPERATIONAL AMPLIFIERS (op-amps). These versatile devices can be used in a wide variety of amplifying, signal processing and waveform-generating applications. An op-amp is basically a low-drift d.c. amplifier with very high open loop differential gain

and good common mode rejection. The symbol for an op-amp together with the pin out data for the 741 IC are shown in Fig. 8.1. There are two input terminals, one called the inverting input (marked $-$); and the other called the non-inverting input (marked $+$). The output, assuming zero offset, will be the difference in signal between the two inputs multiplied by the open loop gain:

$$V_O = A_{vol}\ (V_1 - V_2)$$

Here $(V_1 - V_2)$ is the differential input and, ideally, when $V_1 = V_2$ the output should be zero. In practice some offset always occurs (caused by the slightly different characteristics of the two input transistors). With both inputs held at zero volts, the resulting output caused by the input offset voltage can easily be trimmed to zero by some offset nulling technique. Most IC op-amps such as the 741 are provided with simple facilities, as shown, to achieve this.

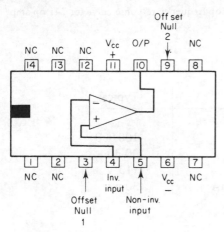

Fig. 8.1A 14-pin n dual-in-line package for the 72741 (top view)

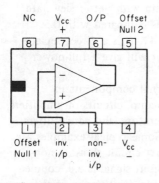

Fig. 8.1B 8-pin p dual-in-line package

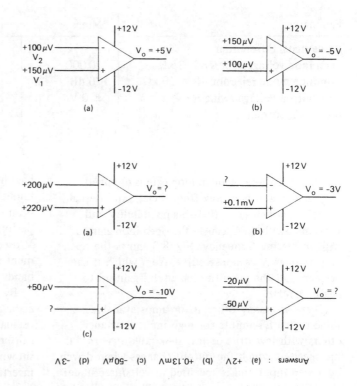

Answers : (a) +2V  (b) +0.13mV  (c) −50μV  (d) −3V

Fig. 8.2. Op-amp exercises

Since the open loop gain $A_{vol}$ is so large (typically 100 dB at d.c. and low frequencies), only a tiny difference in voltage at the input pins is necessary to give a large output.

·   *Note*   100 dB = 100 000 as a voltage ratio.

Take an example with $V_1$ = +150 $\mu V$ and $V_2$ = +100 $\mu V$. Then

$$V_O = 100\ 000\ (150 \times 10^{-6} - 100 \times 10^{-6})$$
$$= +5\ V$$

The output will be at approximately +5 V provided that the power supplied to the IC is at least ± 7 V. If the inputs are now changed, giving $V_1$ = +100 $\mu V$ and $V_2$ = +150 $\mu V$, the new value of output will be

$$V_O = 100\ 000\ (100 \times 10^{-6} - 150 \times 10^{-6})$$
$$= -5\ V$$

The output is inverted.

The above examples are illustrated in Fig. 8.2 together with some exercises. In each case calculate the approximate d.c. output. Assume that any error due to input offset is negligible.

The following extract from the 741 specification gives an indication of the quality of an IC op-amp.

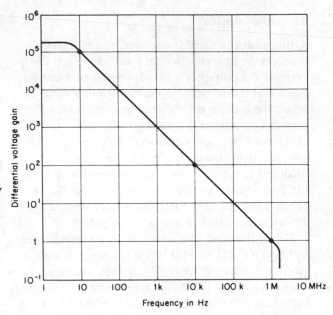

Fig. 8.3 Open loop frequency response curve for 741 op-amp

| Parameter | Min. | Max. | Typical |
|---|---|---|---|
| Large signal differential open loop voltage gain $R_L \geqslant 2$ k$\Omega$ | 20 000 | – | 100 000 |
| Common mode rejection $Rs \leqslant 10$ k$\Omega$ | 70 dB | – | 90 dB |
| Max output voltage swing $R_L \geqslant 2$ k$\Omega$ | ±10 V | – | ±13 V |
| Input bias current | – | 200 nA | 30 nA |

The high value of open loop gain is obtained only at d.c. and very low frequencies. An internal capacitor, which gives the op-amp stability and prevents oscillations, causes the open loop gain to fall with increasing frequency. Fig. 8.3 shows the open loop frequency response curve, from which it can be seen that the amplifier has useful gain up to nearly 1 MHz.

There are many other IC op-amps available, some with FET inputs for high input resistance, others with low drift or fast slew rate. *Slew rate* is the speed at which the output changes when driven by a step input and is specified in volts/$\mu$sec under closed loop unity gain conditions. Slew rate limiting has the effect of turning a sine wave input (of a few kilohertz) into a triangular wave output. The 741s is an example of an op-amp with better slew rate performance, 20 V/$\mu$sec as compared with 0.5 V/$\mu$sec for the standard 741. It is a direct replacement for an 8-pin 741 but will give a full-power bandwidth of 200 kHz.

By using suitable external components in association with an IC op-amp, circuits of amplifiers, oscillators, waveform generators, and active filters can be readily designed. Some typical examples are shown in Fig. 8.4. The first circuit shows a non-inverting amplifier. The input signal is a.c. coupled to pin 3 and $R_2$ sets the input resistance of the

## (a) Pre-amplifier

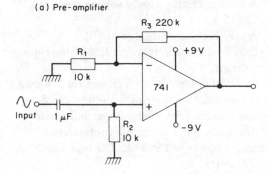

## (b) Square wave oscillator

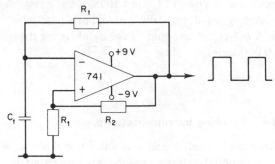

**Fig. 8.4** Typical examples of circuits using op-amps

circuit. A portion of the output signal is fed back to the inverting input to oppose the input signal. The gain is set by $R_1$ and $R_3$ and is given by

$$A_c = \frac{R_3 + R_1}{R_1} = 23$$

The second example shows a 741 used to produce square waves. When the power is first applied, $C_t$ will be uncharged, so the op-amp output will saturate at its positive level ($V^+_{sat}$). A portion of this output voltage is fed back via $R_2$ and $R_1$ to the non-inverting input. The voltage on the non-inverting input will be

$$V_+ = V^+_{sat} \left( \frac{R_1}{R_1 + R_2} \right)$$

As $C_t$ charges via $R_t$ the voltage on the inverting pin rises positive. When this voltage just exceeds the level on the non-inverting input, the op-amp switches rapidly to its negative saturated level $V^-_{sat}$. The

voltage on the non-inverting input also reverses to become negative.

The capacitor $C_t$ now discharges via $R_t$ towards $V^-_{sat}$ until the voltage on the inverting terminal is just more negative than the level set up on the non-inverting input. When this occurs, the op-amp output is again forced to switch to $V^+_{sat}$ and the cycle recommences. In this way the circuit produces continuous square waves. Note that both feedback paths control the frequency since the $R_tC_t$ time constant will determine the charge and discharge rate, while the potential divider $R_2 R_1$ determines the switching points. The frequency is given by the formula:

$$f = \frac{1}{2R_t C_t \log_e [1 + 2R_1/R_2]}$$

The exercises that follow later in this chapter will show applications of op-amps and some other analogue ICs.

## 8.3  Digital ICs

In digital circuits the input data is represented by groups of "highs" or "lows". Using positive logic convention, a high level is represented by a positive voltage and a low level by a voltage at or near zero. Thus the inputs and outputs of digital ICs are signals switched between two well defined states. With TTL, logic 0 is typically 200 mV (not greater than 400 mV) and logic 1 is typically 3.3 V (not less than 2.4 V). The particular advantages of using information in digital rather than analogue form are

(1)  A signal is indicated as either high or low so there is less ambiguity and much less chance of error.
(2)  Digital information can be easily transmitted, stored, and processed without degradation.
(3)  Data can be reshaped after transmission.
(4)  Noise and interference has much less effect.
(5)  Many two-state devices exist.

The last point shows that digital logic is based on two-state devices, i.e. the device is either ON or OFF and gives either a HIGH or a LOW output.

Digital ICs can be grouped into combinational and sequential types. With COMBINATIONAL LOGIC, various input conditions must be met simultaneously to give an output. Therefore combinational logic is made up of gates such as the AND, OR, NAND, NOR, NOT and exclusive-OR.

In SEQUENTIAL LOGIC, the digital elements possess a memory and the resulting output from a sequential logic IC will depend upon the input and the previous state of the circuit. The basic building block of all sequential circuits is the bistable. These can be connected together, often inside an IC, to make counters, shift registers, and memories.

It is not proposed to explain here the theories of digital logic or in great detail how the internal circuit of a digital IC operates (except where required in the exercises). This chapter is intended only as a brief introduction to the subject. However some knowledge of the various digital IC logic families is essential. The earlier types such as RTL (resistor transistor logic) and DTL (diode transistor logic) are now obsolete. However there is much equipment using DTL still in use. The main important logic families are as follows.

### (1) Transistor Transistor Logic (TTL)

A widely used logic type that is available with many functions (look up a manufacturer's catalogue), TTL combines fast speed with moderate power consumption and reasonable levels of noise immunity. It was developed as a successor to DTL and is continually being updated. Later versions use Schottky type transistors which improve the switching speed.

### TTL types

| | | |
|---|---|---|
| Standard | 74 series | |
| Schottky | 74S | |
| Low power Schottky | 74LS | |
| Advanced low power Schottky | 74ALS | |
| High speed | 74H | being |
| Lower power | 74W | phased out |

### (2) Complementary MOS Logic (CMOS)

These are made from combinations of p and n channel enhancement mode MOSFETs. This type of construction results in a number of particular advantages when compared to TTL. The unique features are

(a)  A very low power consumption (about 10 nW/gate under d.c. conditions).

(b)  A wide operating supply voltage range (+3 V to +15 V).

(c)  A very high fan-out (at least 50).

(d)  Excellent noise immunity (45% of the supply voltage).

Therefore CMOS logic is often chosen for low-cost low power consumption systems, especially those for use in an electrically noisy environment and where speed of operation is not the prime consideration. The devices are not as fast as TTL. The type numbers are the 4000 series.

### (3) Emitter Coupled Logic (ECL)

This is more rarely used and does not have the same extensive range as TTL and CMOS. It has a fast operating speed, typically 2 nsec, but a relatively high power consumption. Type numbers are the 10 000 series.

## 8.4 Servicing Instruments Containing ICs

Since the IC itself constitutes a functional block, with a minimum of external components being used, it is obvious that a failure of one part inside the IC will lead to a complete loss of performance and the IC will then have to be replaced. One internal component failure renders the whole IC useless. Naturally ICs are designed to give very high reliability but failures will and do occur. Some of these failures will be caused by the natural environmental stresses acting on the IC, weakening perhaps an internal connecting lead and finally causing an open circuit. Alternatively "spikes" on the supply rails, or large current surges at switch-on, can cause failure at semiconductor junctions. It is the designer's job to ensure that the power supply is well regulated and filtered. TTL logic, for example, has a maximum supply voltage rating of 7 V; no fault must be allowed to occur in the power supply which would put an over-voltage on to the ICs. This would cause many of them to over-heat and burn out.

When testing equipment containing ICs take care not to short pins by using large test probes, avoid the use of excessive heat when unsoldering a component, and *never* remove or plug in an IC to its socket without first switching off the power supply. This is when excessive surge currents can occur, and it is possible for a complete batch of ICs to be destroyed one after another by an unskilled operator.

For fault finding on IC units follow a logical procedure:

(a) Check power supply *at the IC pins*. Is it within its rated value? Is the ripple level low? If the answer is yes, proceed.

(b) Make sure that the required input is present at the IC pin indicated on the diagram.

(c) Check for a suitable output.

(d) Check visually and with a meter for any open or short circuits in the copper track to the IC.

Several aids for servicing ICs are available, such as IC inserters, test clips (reduce risk of accidental shorting), logic probes, etc. Use them wherever possible.

If an IC has to be removed by unsoldering, always use a desoldering tool to remove the solder from each IC pin in turn until the IC can be lifted out. It's worth taking time over this job to avoid damaging the copper track of an expensive printed circuit board.

**Fig. 8.5** Heater control unit using a 741 op-amp

## 8.5   Exercise: Heater Control Unit using a 741 Op-Amp (Fig. 8.5)

In many circuits IC op-amps are used together with other components to give a particular function. The 741, because of its high differential gain and excellent common mode rejection, is ideal for amplifying the difference signal from a bridge circuit. In this example a d.c. bridge is used with a thermistor (GM473) as the sensing element. The signal from the bridge is fed to the inputs of the 741 which is wired as a comparator, with a small amount of positive feedback via $R_4$.

While the temperature inside the enclosure is low, the thermistor has a fairly high resistance (47 k$\Omega$ at 25°C) and the output of the 741 will be high. This level forward biases $Tr_1$ via the zener diode and $R_5$. $Tr_2$ also conducts and takes a current of approximately 0·8 A. The power dissipation of $Tr_2$ is then nearly 10 W, and this heats up the small enclosure. A current limiting circuit is provided by $Tr_3$ since when the output current increases beyond 0·8 A the voltage across $R_8$ rises to about 600 mV and this causes $Tr_3$ to conduct, thus diverting base current from $Tr_1$.

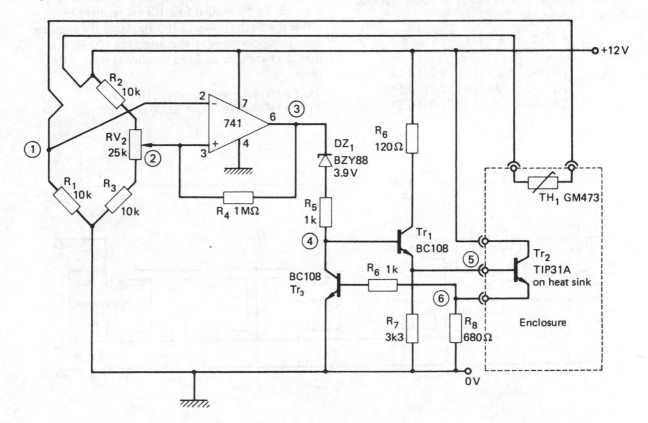

As the temperature in the enclosure rises the resistance of the thermistor falls so that at some point dependent upon the setting of $RV_1$, the output of the 741 falls. This cuts off $Tr_1$ and $Tr_2$ to stabilize the temperature inside the enclosure.

## Questions

(1) State the voltages you would expect to measure with a standard multimeter at all test points when the unit has just been switched on. The temperature in the enclosure is about 25°C. Assume $RV_1$ is at mid-position.

(2) The unit fails so that it will not heat. The voltages measured were as follows. State, with supporting reasons, the possible component failures that could cause this fault.

| TP | 1 | 2 | 3 | 4 | 5 | 6 |
|----|---|---|---|---|---|---|
| MR | 2.1 | 8.1 | 0.26 | 0 | 0 | 0 |

(3) The unit fails so that the temperature rises to a high value and cannot be controlled. After studying the following voltages, state in each case, with a supporting reason, the possible component failures that could cause each fault.

| Fault | 1 | 2 | 3 | 4 | 5 | 6 |
|-------|---|---|---|---|---|---|
| A | 11.1 | 6.2 | 12 | 1.9 | 1.25 | 0.53 |
| B | 11.1 | 6.2 | 2.3 | 1.95 | 1.35 | 0.56 |
| C | 2.1 | 1.3 | 7.6 | 1.9 | 1.25 | 0.53 |

(4) What would be the full symptoms for the following faults?
  (a) $DZ_1$ open circuit.
  (b) $Tr_2$ base emitter short circuit.
  (c) 741 open circuit inverting input.
  (d) $R_1$ open circuit.

## 8.6   Exercise: Frequency Standard Circuit using TTL Logic (Fig. 8.6)

This circuit shows the use of standard TTL logic gates to produce an accurate stable frequency of 100 Hz. A quad-two input positive NAND gate (7400N) is used together with a 1 MHz crystal to provide a 1 MHz square wave oscillation. $P_{1d}$ and $P_{1c}$ are operated as linear amplifiers because of the feedback formed by the resistors $R_2$, $R_1$ and $R_4$, $R_3$. Since positive feedback is provided by the crystal, the circuit oscillates at 1 MHz. $P_{1b}$ is used to buffer the 1 MHz output signal from the crystal. In the same way $P_{1a}$ will transmit the 1 MHz signal to the decade counter $P_2$ when the "gate" input is at logic 1.

Each of the decade counters (7490N) divide the input by 10, so that four ICs are required to divide the 1 MHz down to 100 Hz. Since the counters operate only on logic level changes the 100 Hz will have the same accuracy and stability as the 1 MHz oscillator.

**Fig. 8.6** Frequency standard circuit (100 Hz)
$P_1$ is an 7400N quad 2 input NAND gate

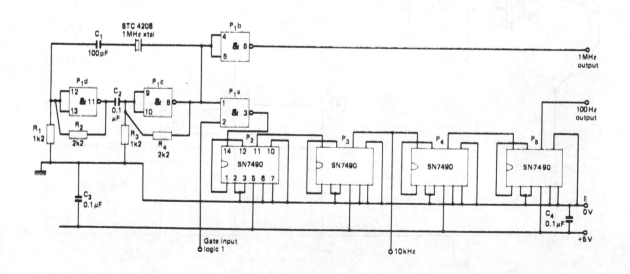

The following definitions apply to logic measurements in TTL logic:

Logical 0 voltage: less than +400 mV (typically +200 mV)

Logical 1 voltage: greater than +2·4 V (typically +3·3 V)

N.B. These voltages are measured with respect to the 0 V rail.

As stated previously, before checking the operation of a particular IC, always verify the presence of the +5 V and 0 V supplies. Make the measurement directly at the appropriate IC pins, *not* between board connections or on the printed circuit wiring.

Never check the voltage levels on an IC by connecting the −ve meter lead to chassis and monitoring the voltage with the +ve lead, since a break in the ground line to the IC will not be indicated.

Most logic circuits, especially those of the series type as in this example, are relatively easy to fault find since the symptoms indicate the fault. For example suppose the unit fails to give a 100 Hz output. After checking the presence of the +5 V line, one should then check the output from the 1 MHz oscillator, then the input and output of ICs $P_2$ to $P_3$ until the fault is located.

## Questions

(1) The unit fails so that the 1 MHz output remains at +3·3 V. The 100 Hz output is present. State the portion of the circuit that has failed.

(2) What is the purpose of $C_3$ and $C_4$? Where should they be located?

(3) The unit has a fault such that there is no 100 Hz or 10 kHz output, although the 1 MHz output is correct. Which ICs could be at fault, and how could you quickly locate the fault?

(4) How could the accuracy and stability of the 100 Hz output be checked in the laboratory?

(5) What would be the result of the following faults?

(*a*) $P_4$ open circuit track to +5 V line.

(*b*) $C_1$ open circuit.

(*c*) $P_{1a}$ open circuit output.

(6) How could the operation of each decade be checked, assuming that the oscillator has failed?

## 8.7 Exercise: Power Unit (Fig. 8.7)

Since the 741 is a differential amplifier with very high gain it makes an ideal comparator and error amplifier for a linear series stabiliser. This gives a relatively simple circuit with quite a good performance. A circuit example is given in Fig. 8.7. The specification is as follows:

| | |
|---|---|
| Output voltage range | 9 V |
| Max. output current | 0·4 A (current limited) |
| Ripple | 2·5 mV pk-pk. |
| Load regulation | Better than 0·02 per cent zero to full load |
| Line regulation | ±10 per cent change in mains gives less than ±0·05 per cent change in output |

The circuit is a conventional stabiliser, the non-inverting input (+) of the 741 amplifier being held at a constant voltage by the Zener diode (5·6 V). The inverting input (−) is taken to a potentiometer. Since the 741 has a high gain (100 000) it only requires a difference of a millivolt or so between the (+) and (−) input terminals for the output to be driven positive or negative by a large amount. If, for example, the input difference is 0·1 mV negative, the output of the op-amp would try to move several volts positive. The output therefore assumes a voltage which will cause the difference between the Zener voltage and the voltage on $RV_1$ slider to be as small as possible.

Neglecting the action of the current limit we can see how the circuit operates to hold the output constant by imagining a fall in output caused by an increased load. This would provide at the inverting input of $IC_1$ a net negative input. Pin 6 will go positive causing $Tr_1$ (the series element) to conduct more, thus forcing the output back to very nearly its initial value. The opposite will occur if the output rises for any reason.

The changes in output voltage from zero to full load current are very small because of the very high gain of the 741. Thus one IC gives this relatively simple power supply very good performance.

The maximum output current is limited to about 0·4 A. If the current increases beyond this the voltage across $R_3$ causes $Tr_2$ to conduct and the output voltage falls. Thus, if $Tr_1$ is mounted on a small heat sink no damage occurs if the output is accidentally short circuited.

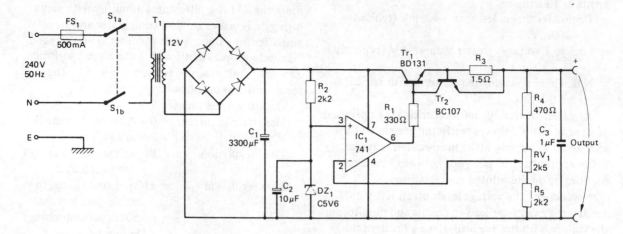

**Fig. 8.7** Power supply unit

To start, ignore all the other components and concentrate on possible faults with the 741 IC. It is possible for many different faults to occur inside the actual silicon chip and in the connecting leads. Internal shorts or opens may occur, the connecting pins or track can become open circuit or short to adjacent pins. Naturally it is not always possible to pinpoint the actual fault but it is a good idea to sort out the type of fault since it may show a possible external cause.

Take for example an internal open circuit on the inverting input of the 741. The voltage readings with a 100 mA load are:

| Pin No. | 2 | 3 | 7 | 6 | Output |
|---------|------|-----|------|------|--------|
| Voltage | +9·3 | 5·7 | 16·2 | 15·1 | 12·5 |

The symptoms are no control and poor regulation.

Since the inverting lead is open circuit, the output of the 741 has been driven hard positive, forcing the output to rise. $RV_1$ will have no control. Note that there is an excessive positive difference signal between 2 and 3 which should drive the output down, not up.

If the output going high is a symptom for an open circuit pin 2, then the reverse effect can be expected if pin 3 were open circuit. This is in fact the case as indicated:

| Pin No. | 2 | 3 | 7 | 6 | |
|---------|-----|-----|------|-----|-----------|
| Voltage | 1·7 | 5·7 | 16·2 | 3·5 | No control |

It is important to note that similar symptoms would be produced if the Zener or $C_2$ became short circuit, or if $R_1$ went open, except for the fact that pin 3 would then read zero volts.

**Questions**

(1)  What fault on the IC would give the following:

| Pin No. | 2 | 3 | 7 | 6 | Output |
|---------|---|-----|------|---|--------|
| Voltage | 0 | 5·7 | 16·2 | 0 | 0 |

(2)  State the symptoms for the following faults:

(a)  Open circuit lead to pin 7.
(b)  $RV_1$ slider open circuit.
(c)  $R_5$ open circuit.
(d)  $R_2$ open circuit.

## 8.8   Exercise: Gated Pulse Generator (Fig. 8.8)

The TTL logic IC 7413 contains two Schmitt trigger circuits. Each Schmitt gate has 4 inputs and a NAND output; all 4 inputs must be high to give a low output. If any one input is low, the output will be high. A feature of a Schmitt gate is the hysteresis between the positive going threshold ($V_{T+}$) of 1.7 V and the negative going threshold ($V_{T-}$) of 0.9 V. It is this 800 mV hysteresis that enables the gate to be used as a pulse generator. In gate A the output is connected back via $R_1$ and the emitter follower $Tr_1$ to one gate A input. (Note that the other three inputs are connected to +5 V.) When the output of A is high, $C_1$ charges up causing the voltage on the input to rise. When this voltage on gate A input exceeds $V_{T+}$ (1.7 V), the output switches low. $C_1$ now discharges, via $R_1$ and the gate output, towards zero volts. When gate A input voltage falls below $V_{T-}$ (0.9 V), the output of the gate again switches to a high state and the next cycle starts. In this way the output of gate A is a continuous pulse waveform with a frequency of about 1 kHz. The waveform has a mark-to-space ratio of about 1.2:1.

The output of Schmitt gate A is directly connected to one of the input pins of Schmitt B. Gate B is also wired as a pulse generator with frequency set to about 20 kHz by $C_2$ and $R_3$. When the output of gate A is high, the pulse generator formed by gate B is free to oscillate, but B will be inhibited when gate A is low. Thus the output of B is a waveform consisting of "bursts" of 20 kHz pulses. The waveforms can be checked using a CRO.

### Questions

(1)   Sketch the waveforms you would expect to measure at points (x), (y) and (z).

(2)   State the component (or components) fault that would cause the following symptoms:

(a)   Output (y) an inverted version of (x), i.e. output (y) oscillating at 1 kHz only.

(b)   Output (y) 1 kHz bursts of high frequency pulses (at approximately 10 MHz).

(c)   Continuous train of pulses at 20 kHz from (y). Point (x) at logic 1.

(d)   Continuous high-frequency oscillations ($\simeq$ 10 MHz) from both (x) and (y).

(3)   Explain the symptoms that would occur for the following faults

(a)   $Tr_1$ base/emitter short circuit.

(b)   $Tr_1$ collector/emitter short circuit.

(c)   Open circuit track from gate A to gate B input.

(d)   Gate A output "stuck at 0".

(e)   Gate B output "stuck at 1".

(4)   Suggest modifications to convert the circuit to a bleeping audible alarm.

**Fig. 8.8** Gated pulse generator

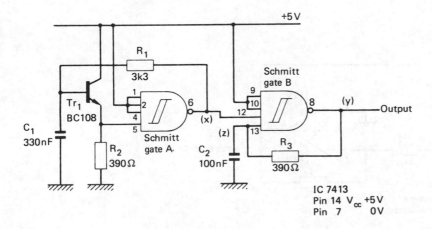

IC 7413
Pin 14 $V_{cc}$ +5 V
Pin  7      0 V

## 8.9   Exercise: Triangle/Square Wave Generator (Fig. 8.9)

Two op-amps are used in this circuit to give simultaneous triangle and square wave output signals. $IC_1$ is wired as an integrator with the time constant set by $R_1$ and $C_1$. An integrator is a circuit that will give a ramp-type output for a fixed input voltage. In the circuit $IC_2$ is used as a comparator to give an output that switches between $+V_{O(sat)}$, and $-V_{O(sat)}$, where $V_{O(sat)}$ is the saturated output level of the op-amp (approximately 8 V). This plus or minus 8 V level is fed back to the integrator's input to cause it to ramp in alternate directions. Suppose, at switch on, that the output of $IC_2$ goes positive to $+V_{O(sat)}$. This positive level is fed back to the integrator causing a fixed current to flow in $R_1$. This current also flows through $C_1$ forcing the output of $IC_1$ to ramp in a negative direction. When the ramp reaches approximately –4 V, the voltage of pin 3 of $IC_2$ just moves negative. When this occurs, $IC_2$ output switches negative to $-V_{O(sat)}$ (–8 V). This reverses the flow of current in $R_1$ and $C_1$, and $IC_1$ output now ramps in a positive direction. When the ramp reaches about +4 V, the comparator again switches and the next cycle begins. In this way the circuit produces continuous triangle waves at $IC_1$ output and square waves at $IC_2$ output as shown in Fig. 8.10. The frequency of operation depends on the time constant of the integrator and also the values of $R_2$ and $R_3$. These two resistors determine the amplitude of the triangle wave. The frequency of oscillation is given by

$$f = \frac{1}{4C_1 R_1 \ (R_2 / R_3)}$$

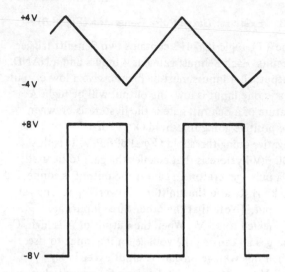

Fig. 8.10 Waveforms in triangle generator

## Questions

(1)   Calculate the frequency of oscillation.

(2)   A fault develops causing the circuit to oscillate at almost 125 kHz with both waveforms distorted. State the component fault.

(3)   State the symptoms for the following faults:
    (a)   $R_1$ open circuit
    (b)   $C_1$ short circuit
    (c)   $R_2$ open circuit

(4)   The circuit does not oscillate and both IC outputs are high at nearly +8 V. State the component(s) or connection fault that could cause these symptoms.

Fig. 8.9 Triangle/square wave generator
($IC_1$ and $IC_2$ are 741 types)

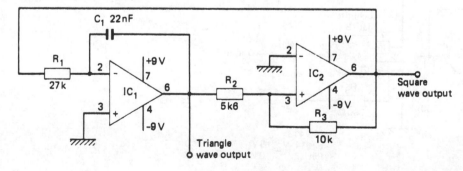

## 8.10 Exercise: Interface Circuit (Fig. 8.11)

This small circuit, using two CMOS ICs, is part of the interface between a microcomputer and a cassette recorder. When the control input is high, at logic 1, the output will be 2·4 kHz square waves. With the control input low, at logic 0, the output will be 1·2 kHz square waves. In this way, a high and a low audio tone can be generated for recording purposes.

The clock input at 4·8 kHz is divided by the two D-type bistables contained in the 4013 IC to provide the 2·4 kHz and 1·2 kHz square waves. The other IC, a quad 2-input NAND type 4011, performs the gating function. When the control input is high, gate A output will be held low, thus closing gate C. Gate B, however, will be open and will transmit the 2·4 kHz square waves to the output via gate D. The reverse conditions are set up when the control switches to logic 0, so that gate B is off and gate C on to give 1·2 kHz square waves at the output.

## Questions

(1)    A failure occurs in which only 1·2 kHz square waves can be obtained. When the control input is logic 1, the output remains steady at logic 0. State the portion of the circuit that is at fault.

(2)    What output will result if the control lead is removed?

(3)    The unit fails with the following symptoms:

| Control input | Output |
|---|---|
| 1 | 2·4 kHz square waves |
| 0 | 0 |

State the portion of the circuit that is at fault and the type of fault.

(4)    Describe the symptoms for the following faults:
   (a)    Gate B output "stuck at 0."
   (b)    An open circuit track to the clock input of the second bistable.

(5)    Which section of the circuit would be at fault to give the following symptoms?

| Control input | Output |
|---|---|
| 1 | 1 |
| 0 | 0 |

**Fig. 8.11** Interface circuit

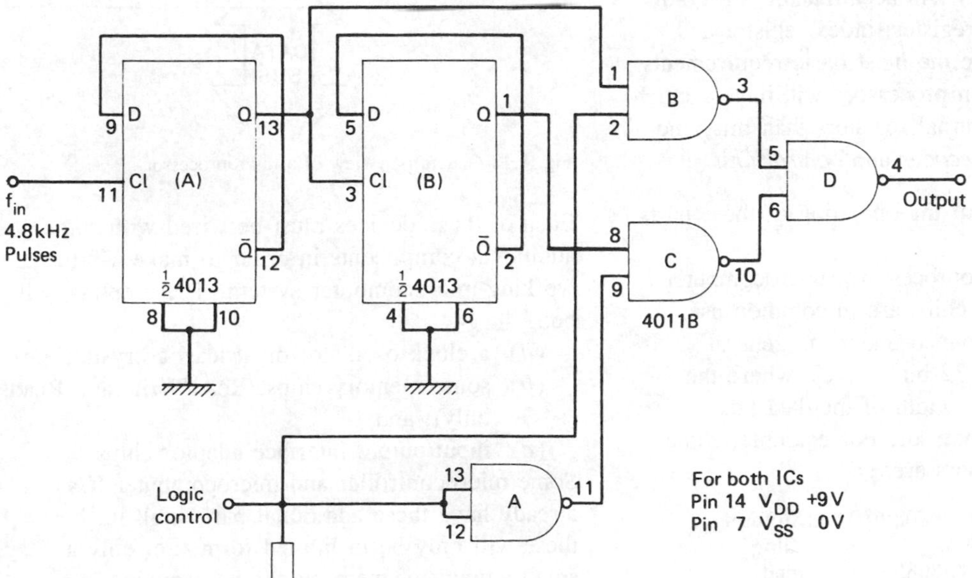

$f_{in}$
4.8 kHz
Pulses

$\frac{1}{2}$ 4013

$\frac{1}{2}$ 4013

4011B

Logic control

For both ICs
Pin 14  $V_{DD}$  +9 V
Pin  7  $V_{SS}$   0 V

Output

## 8.11 Fault finding on microprocessor based systems

The microprocessor has obviously been a most important recent development in electronics. In some cases these ICs have replaced traditional hardwired digital logic whilst in others they have opened up entirely new application areas. There are now large numbers of microprocessor systems in use, all of which will from time to time require maintenance and repair. These notes are intended as a basic introduction to fault finding on micro-processor based systems; but since the subject is quite extensive some further study of other texts will be essential.

A microprocessor is defined as a very large scale integrated circuit designed to act as the central processing unit (CPU) of a digital computer. This means it will consist of the following main sections (see Fig. 8.12):

(1) an *arithmetic and logic unit* (ALU), the portion that performs operations such as adding, subtracting, anding, comparing, and so on;

(2) a set of *registers*. An accumulator, program counter, status register, index register and stack pointer are the most basic requirements but most modern processors will have a much larger set of internal registers than this; and

(3) an *instruction decoder* and *control unit*.

A brief description of the operation of these parts will be given later.

A variety of microprocessors, microcomputer and microcontroller chips are in common use. They are usually grouped under the general heading of 8, 16 or 32 bit devices, where the number indicates the width of the data bus employed by the processor. For example, some typical microprocessors are:

| 8 bit machines: | 6800/6802/6809 | Motorola |
| | Z80 | Zilog |
| | 8080/8085 | Intel |
| | 6502/65C02 | MOS Technology |
| 16 bit machines: | Z8001/2 | Zilog |
| | 8086/8088 | Intel |
| | 68000 | Motorola |
| | 80286/80386 | Intel |

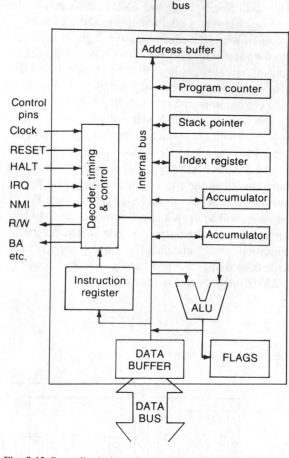

Fig. 8.12 Generalised view of a microprocessor

Each of these devices must be wired with some additional components in order to make a full working microcomputer system. These extras will be:

(*a*) a clock oscillator or at least a crystal;

(*b*) some memory chips (Read/Write and Read only); and

(*c*) input/output interface adaptor chips.

Some microcontroller and microcomputer ICs already have these additional parts built in, but these will only be in limited form, i.e. only a small amount of memory can be provided.

A basic system for a microcomputer is shown in Fig. 8.13 where it can be seen that communication between the various ICs is by three highways each called a BUS. A bus is therefore a group of

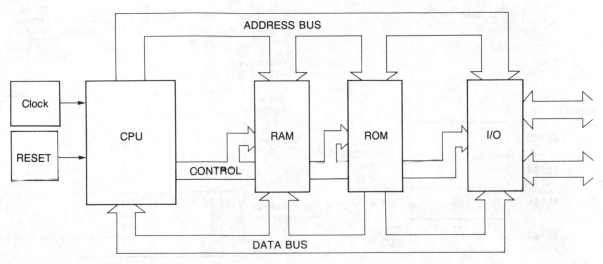

**Fig. 8.13** Basic microcomputer system

copper tracks on a p.c.b. or a set of wires between boards, that are linked for a common purpose. The *address bus* is used by the microprocessor to call up locations in memory (RAM or ROM) or I/O, the *data bus* to transfer data or instructions between the memory and the microprocessor and the *control bus*, consisting of a set of control lines such as read/write, interrupt, valid memory address and so on, used to synchronise the actions that the system is required to make.

A microprocessor with its additional chips forms an extremely flexible and powerful digital system which can be used for computing or control or as a completely dedicated system. It achieves its flexibility because it can be reprogrammed to carry out an almost unlimited variety of tasks. A set of instructions (the PROGRAM) is held in memory and these instructions will be fetched, decoded and executed in sequence by the microprocessor until the task is complete. For example, a segment of an *assembly language program* which reads some switches from an input port and then drives a motor connected to an output port when all three input switches are high would, in 6809 code, be written as:

```
Read:  LDA PORTA  ; read switches
       CMPA #03    ; all switches are set?
       BNE Read    ; if not repeat read
```

LDB #$FF    ; otherwise load B with 1's
STB PORTB   ; and output to motor
RTS         ; return from subroutine

Each line of this assembly language program contains a *mnemonic* such as LDA, BNE, STB, etc., all shorthand ways of writing an instruction. Thus LDA PORTA means LOAD ACCUMULATOR *A* FROM AN I/O ADDRESS LABELLED PORTA. Note also that $ means hexadecimal number and that # means immediate.

Such an assembly language program (called the software) is converted into machine code and machine language by more software run on another computing system. This other software is called an ASSEMBLER. The final program is then downloaded into the target machine, i.e. the system that is to operate the program, where the instructions, addresses, and the data for the machine must then all be in binary form.

In any computing system the various levels of language which can be used can therefore be listed as:

(1) High level language — programming languages such as PASCAL, C, ADA, FORTRAN, etc.

(2) Assembly language — mnemonic code specific to the processor used

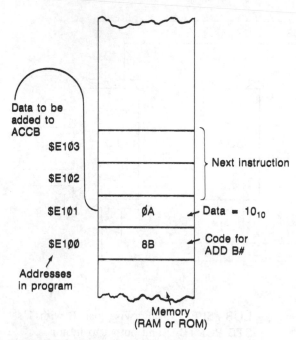

**Fig. 8.14** Illustration of immediate addressing

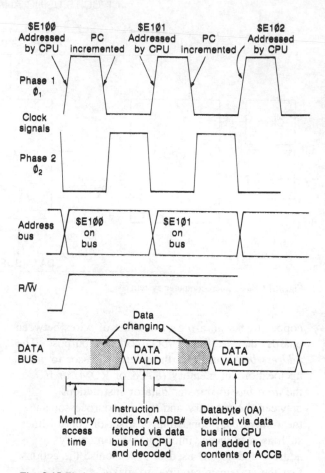

**Fig. 8.15** Timing diagram

(3) Machine code — mnemonics, addresses and data in hexadecimal code, i.e. in numbers to base 16

(4) Machine language — instructions, addresses and data in binary number code

In fault finding procedures it is the lower level languages that are more likely to be used. But it is important to realise that these will be different for each type of processor – each has its own unique set of instructions and codes. Let us consider a simple instruction such as adding the number 10 to a register. In 6809 assembly language this instruction is written:

ADD B #10 ; add decimal 10 to register B
which in machine code (hexadecimal) is:
8B 0A
and finally in machine language (binary) becomes:
10001011 00001010

Quite often fault finding may require the engineer to examine portions of code either held in memory or by running through a sequence. Special aids, such as the *Logic Analyser*, are then used. These instruments are designed to capture a portion of a program and display the results as either binary or

hexadecimal code, or as a series of time related waveforms. Another useful feature is called *disassembly*, where the code taken from the system under test and held in the Logic Analyser's memory can be converted back into its mnemonic form.

Let us now consider the actions required by a 6809 system to complete the instruction ADD B #10. Assume the instruction is held at address $E100 with the data following immediately at address $E101 (see Fig. 8.14). All events in the 6809 are synchronised by a two phase non-overlapping clock and in Fig. 8.15 an indication of the timing is shown. The sequence of events is as follows:

(1) The binary pattern for the hex. address $E100 (this is 1110000100000000) is placed on the address bus by the microprocessor.

The program counter inside the processor is then incremented ready for the next program address and the read/write line set high to switch the memory to read mode.

(2) The contents of address $E100, the hex. code 8B equivalent to the instruction ADD *next number in program memory to register B*, is fetched from the memory location into the instruction register of the micro-processor. The instruction is decoded and internal logic set as required.

(3) The next address, $E101, is placed on the address bus and the program counter is again incremented.

(4) The contents of $E101, *the data byte* $0A($10_{10}$), is fetched into the micro-processor via the data bus.

(5) The ALU inside the microprocessor has already been switched to parallel add mode and the contents of register B are now

added to the data byte $0A. The result of the addition is then returned to register B with any carry setting the C flag in the condition code register (status register).

The instruction execution is complete. Two clock cycles have been used in its fetch and execution. The addressing mode used in this case is called IMMEDIATE. An *addressing mode* refers to the way in which data is obtained for the processor by the code of the instruction. In any processor family there will be a variety of addressing modes. In the 6809, for example, the data byte $0A could also be obtained from the following sources.

(*a*) From anywhere in memory using the *EXTENDED ADDRESSING* mode, assuming the data is held at address $E210,
ADDB $E210 ; add the contents of address $E210 to register B (see Fig. 8.16A).

(*b*) From an address pointed to by an index register using the *INDEXED ADDRESSING* mode
ADDB 0,X ; add the contents of the address pointed to by the index register to register B (see Fig. 8.16B).

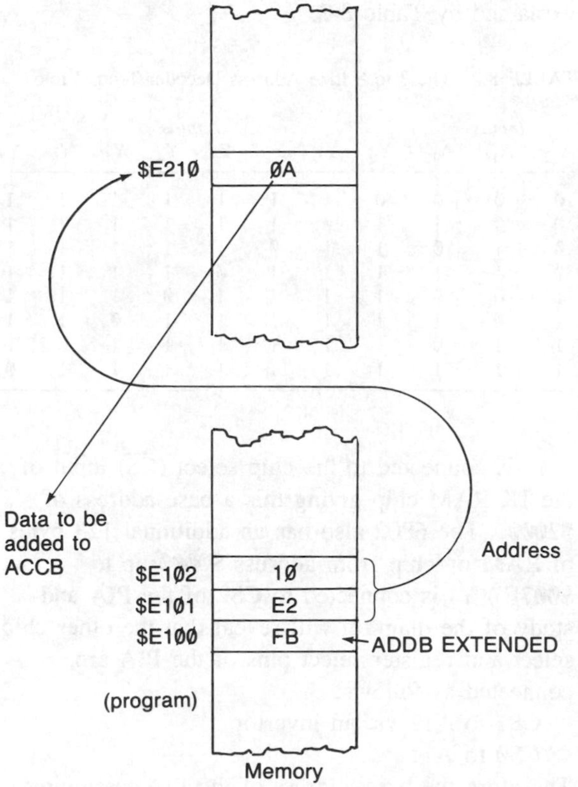

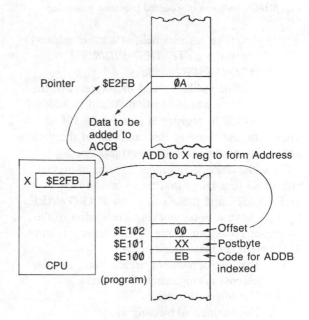

**Fig. 8.16A** Illustration of 6809 extended addressing

**Fig. 8.16B** Indexed addressing in the 6809

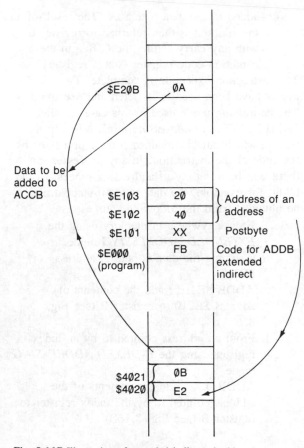

**Fig. 8.16C** Illustration of extended indirected addressing

(*c*) From an address held at another address using the *EXTENDED INDIRECT ADDRESSING* mode
ADDB ($4020) ; add the contents of the address whose location is held in address $4020 to register B (see Fig. 8.16C).

These are just some of the examples of the various addressing modes used by microprocessors.

Since the Service Engineer has the additional task of sorting faults into those caused by SOFTWARE and those caused by HARDWARE, he must have a good working knowledge of the processor contained in the system he is repairing. This entails an understanding of:

(1) The microprocessor architecture, i.e. the internal arrangement of registers.
(2) The instruction set.
(3) The various addressing modes.

This, at first sight, rather formidable task is best tackled by gaining a thorough understanding of one type of microprocessor. When the operation of this is fully absorbed it then becomes a relatively easy job to extend the study to other types.

Consider a basic microprocessor system shown in Fig. 8.17. This is based around a 6802 processor which is an excellent starting point for these studies since its architecture and operation are fairly straightforward and therefore more easily understood. The internal architecture is shown in Fig. 8.17 and consists of the standard registers and portions already mentioned. (For a fuller description of the operation refer to other texts such as *The Microprocessor Sourcebook* by G.C. Loveday, published by Pitman.)

The system consists of the 6802, 1K RAM for data, a 2K EPROM (type 2716) to hold the main operating program and a 6821 PIA for I/O. A 3 to 8 line decoder is used to decode the three most significant address lines $A_{13}$, $A_{14}$ and $A_{15}$. This chip has 8 outputs, only one of which goes low for a unique combination of inputs. This is best explained by Table 8.2.

TABLE 8.2  The 3 to 8 Line Address Decoder Truth Table

| Inputs | | | Outputs | | | | | | | |
|--------|--------|--------|--------|--------|--------|--------|--------|--------|--------|--------|
| $A_{15}$ | $A_{14}$ | $A_{13}$ | $Y_0$ | $Y_1$ | $Y_2$ | $Y_3$ | $Y_4$ | $Y_5$ | $Y_6$ | $Y_7$ |
| Ø | Ø | Ø | Ø | 1 | 1 | 1 | 1 | 1 | 1 | 1 |
| Ø | Ø | 1 | 1 | Ø | 1 | 1 | 1 | 1 | 1 | 1 |
| Ø | 1 | Ø | 1 | 1 | Ø | 1 | 1 | 1 | 1 | 1 |
| Ø | 1 | 1 | 1 | 1 | 1 | Ø | 1 | 1 | 1 | 1 |
| 1 | Ø | Ø | 1 | 1 | 1 | 1 | Ø | 1 | 1 | 1 |
| 1 | Ø | 1 | 1 | 1 | 1 | 1 | 1 | Ø | 1 | 1 |
| 1 | 1 | Ø | 1 | 1 | 1 | 1 | 1 | 1 | Ø | 1 |
| 1 | 1 | 1 | 1 | 1 | 1 | 1 | 1 | 1 | 1 | Ø |

$Y_1$ is connected to the chip select ($\overline{CS}$) input of the 1K RAM chip giving this a base address of $2000. (The 6802 also has an additional 128 bytes of RAM on-chip from address $0000 up to $007F.) $Y_2$ is connected to $\overline{CS2}$ of the PIA and study of the diagram will reveal that the other chip select and register select pins of the PIA are connected as follows:

CS1 to A12 via an invertor
CS0 to A11

Therefore the base address of the PIA, assuming undecoded address lines (X) are zero, is:

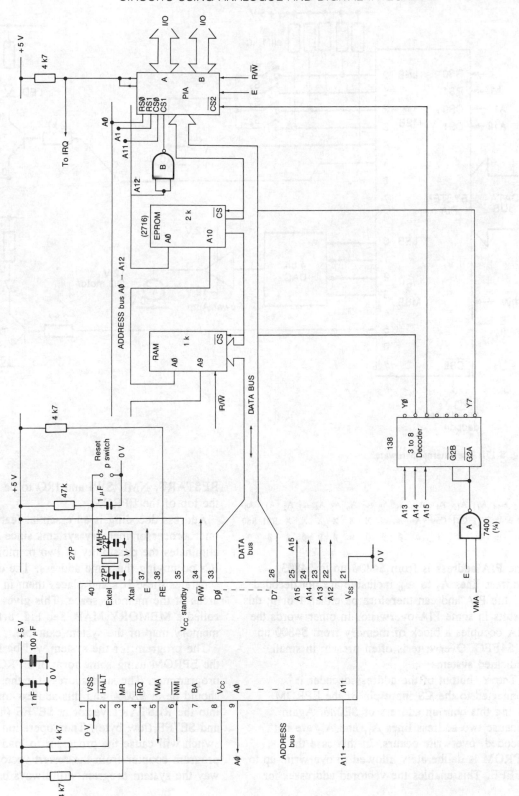

**Fig. 8.17A** The microsystem

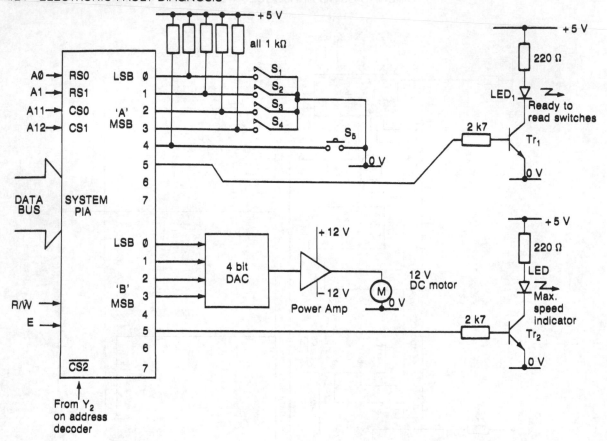

**Fig. 8.17B** The external hardware

| $A_{15}$ | $A_{14}$ | $A_{13}$ | $A_{12}$ | $A_{11}$ | $A_{10}$ | $A_9$ | $A_8$ | $A_7$ | $A_6$ | $A_5$ | $A_4$ | $A_3$ | $A_2$ | $A_1$ | $A_0$ |
|------|------|------|------|------|------|------|------|------|------|------|------|------|------|------|------|
| (Decoder $Y_2$) | | $\overline{CS1}$ | $\overline{CS\emptyset}$ | X | X | X | X | X | X | X | X | X | X | RS1 | RS0 |
| $\emptyset$ | 1 | $\emptyset$ | $\emptyset$ | 1 | $\emptyset$ | $\emptyset$ | $\emptyset$ | $\emptyset$ | $\emptyset$ | $\emptyset$ | $\emptyset$ | $\emptyset$ | $\emptyset$ | $\emptyset$ | $\emptyset$ |

The PIA address is from \$4800 up to \$4803. Address lines $A_2$ to $A_{10}$ inclusive are not decoded for the PIA and can therefore be either 1 or 0; this results in some PIA overwrite. In other words the PIA occupies a block of memory from \$4800 up to \$4FFC. Overwrite is often present in small dedicated systems.

The $Y_7$ output of the address decoder is connected to the $\overline{CS}$ input pin of the EPROM giving this chip an address of \$E000. Again, because two address lines $A_{11}$ and $A_{12}$ are not decoded, overwrite occurs. In this case the EPROM is deliberately allowed to overwrite up to \$FFFF. This enables the vectored addresses for RESTART, NMI, SWI and IRQ to be located at the top of the EPROM.

Address decoding is an essential feature of microprocessor memory systems since its use eliminates the possibility of two or more memory ICs occupying the same address. The decoding separates these ICs and places them in suitable areas of the memory space. This gives what is called a MEMORY MAP. See Fig. 8.18 for the memory map of the system example.

The program for the system will be loaded into the EPROM using some form of EPROM programmer. The start address for the program should be at \$E000 and this address must be set into the RESTART vector at \$E7FE (high byte) and \$E7FF (low byte). Then operation of the reset switch will cause the processor to load the program counter from the restart vector. In this way the system program will always be loaded

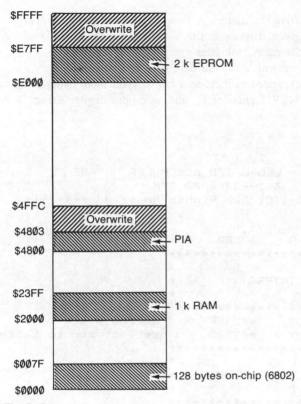

**Fig. 8.18** System memory map

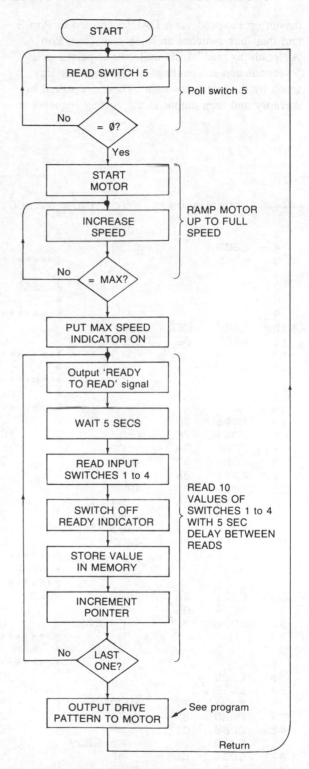

**Fig. 8.19** Flowchart for program

correctly when the system is switched on at power-up or when it is reset.

The program itself must contain instructions to:

(a) set up the stack by loading the stack pointer with a suitable RAM address. In this example a suitable stack area would be $23FF at the top of the 1K RAM. The *stack* is a defined area of RAM used for temporary data storage and for the important tasks of holding return addresses when subroutines are called and processor register contents during an interrupt; interrupt;

(b) set any memory pointers which may be required for tables held in memory;

(c) initialise the PIA for input/output conditions; and

(d) carry out the control task, i.e. read inputs and drive outputs as required.

A typical assembly language program is shown in Fig. 8.19 where it is assumed that a motor is driven up to speed via a DAC connected to Port B and that four switches are then read from Port A. A "ready to read" LED indicator is pulsed every 5 seconds and ten readings of the switches are taken by the microprocessor. These are stored in memory and then output in the correct sequence to drive the motor. A flowchart of the action is also given. However, this is only a simple example of the many tasks the system can be made to perform. Also the hardware can be readily expanded to include a hex-keypad input using a 74C922 encoder IC and an output display using

```
TEST                                    LLOYD I/O ASSEMBLER    PAGE    1
                                        4-19-87 6800 CPM
++++++++ CRASMB V5.2    (C) 1984 by LLOYD I/O, All Rights Reserved ++++++++

    3
    4 * E000                                    ORG     $E000
    5                                 ***************************
    6                                 *                         *
    7                                 *    SET POINTERS          *
    8                                 *                         *
    9                                 ***************************
   10   E000   8E    23FF                        LDS     #$23FF      define stack
   11   E003   CE    2200                        LDX     #$2200      set pointer to table
   12                                 ***************************
   13                                 *                         *
   14                                 *  INITIALISATION OF       *
   15                                 *     THE PIA              *
   16                                 ***************************
   17   E006   7F    4801                        CLR     $4801       clear control reg a
   18   E009   7F    4803                        CLR     $4803       clear control reg b
   19   E00C   86    E0                          LDAA    #$E0
   20   E00E   B7    4800                        STAA    $4800       set I/O SIDE A
   21   E011   86    FF                          LDAA    #$FF
   22   E013   B7    4802                        STAA    $4802       set SIDE B outputs
   23   E016   86    04                          LDAA    #4
   24   E018   B7    4801                        STAA    $4801       reset control a
   25   E01B   B7    4803                        STAA    $4803       reset control b
   26   E01E   7F    4802                        CLR     $4802       motor off
   27                                 ***************************
   28                                 * READ SWITCH S5          *
   29                                 ***************************
   30   E021   B6    4800             READ       LDAA    $4800       read porta
   31   E024   85    10                          BITA    #$10        test bit 4
   32   E026   26    F9 E021                     BNE     READ        return if not zero
   33                                 ***************************
   34                                 *   RAMP MOTOR            *
   35                                 ***************************
   36   E028   5F                                CLRB
   37   E029   5C                     RAMP       INCB                add 1 to reg b
   38   E02A   F7    4802                        STAB    $4802       output to motor
   39   E02D   BD    E074                        JSR     WAIT1       wait .33 sec
   40   E030   C1    0F                          CMPB    #$0F        max speed?
   41   E032   26    F5 E029                     BNE     RAMP
   42   E034   C6    4F                          LDAB    #$4F
   43   E036   F7    4802                        STAB    $4802       max speed led on
```

```
44
45                                  ***************************
46                                  *    READ SWITCHES        *
47                                  ***************************
48       E039    86      20    AGAIN    LDAA    #$20
49       E03B    B7    4800             STAA    $4800        put ready led on
50       E03E    BD    E081             JSR     WAIT2        wait 5 sec
51       E041    7F    4800             CLR     $4800        switch led off
52       E044    BD    E074             JSR     WAIT1
53       E047    B6    4800             LDAA    $4800        read input
54       E04A    84      0F             ANDA    #$0F         mask top 4 bits
55       E04C    A7    00               STAA    0,X          store reading
56       E04E    08                     INX                  move pointer up
57       E04F    8C    220A             CPX     #$220A       ten done?
58       E052    26    E5 E039          BNE     AGAIN
59       E054    CE    2200             LDX     #$2200       restore pointer
60                                  ***************************
61                                  * OUTPUT DRIVE PATTERN    *
62                                  ***************************
63       E057    E6    00     GETNEW   LDAB    0,X          load reg b from table

64       E059    F7    4802             STAB    $4802        output to motor
65       E05C    C1      0F             CMPB    #$0F         is it max?
66       E05E    26    05 E065          BNE     SKIP
67       E060    C6      4F             LDAB    #$4F
68       E062    F7    4802             STAB    $4802        max led on
69       E065    BD    E081    SKIP     JSR     WAIT2
70       E068    08                     INX
71       E069    8C    220A             CPX     #$220A       all ten done?
72       E06C    26    E9 E057          BNE     GETNEW
73       E06E    CE    2200             LDX     #$2200       restore pointer
74       E071    7E    E021             JMP     READ         return to poll S5
75                                  ***************************
76                                  *    SUBROUTINES          *
77                                  ***************************
78       E074    FF    2300    WAIT1    STX     $2300        save pointer
79       E077    CE    FFFF             LDX     #$FFFF       load multiplier
80       E07A    09             LOOP1   DEX
81       E07B    26    FD E07A          BNE     LOOP1
82       E07D    FE    2300             LDX     $2300        restore pointer
83       E080    39                     RTS                  return
84       E081    36             WAIT2   PSHA                 save reg a
85       E082    86      0F             LDAA    #$0F         load a with 15
86       E084    BD    E074    LOOP2    JSR     WAIT1
87       E087    4A                     DECA
88       E088    26    FA E084          BNE     LOOP2
89       E08A    32                     PULA                 restore a
90       E08B    39                     RTS
91       E08C                   END
```

some multiplexed 7-segment indicators. Another PIA, located at say $8000, and suitable buffer/drivers would be required.

As stated previously, faults in microprocessor systems can be caused by either software or hardware. In many cases it is possible for the Service Engineer to remove the system FIRMWARE held in its ROM and insert another EPROM with a known working piece of test software. Such software would have to initialise the system as described above but could then be used to test RAM by writing $FF to each location and reading back. In a similar way I/O function could be tested. Not only does this eliminate a software fault, if it exists, but it also allows tests to be made on all the major portions of the circuit. Hardware faults can be caused by:

(a) the microprocessor itself;
(b) the power supply to the system;
(c) the decoding and bus system;
(d) the memory chips;
(e) any I/O chips; or
(f) interfacing and peripheral hardware.

Although microprocessors are complex dynamic pieces of electronics it is possible to isolate and locate faults. Consider a fault such as the 138 decoder $Y_7$ output stuck at "1". In this case the 2K EPROM would never be enabled via its chip select since it requires a low on this pin. No programs at all could be run on the system.

In a similar way if an address bus fault such as $A_{12}$ going open circuit occurred it would not be possible for the system to address the PIA. An open circuit line floats high, which in this case would set the PIA CS1 input permanently low.

It is important to try and isolate any hardware fault to one part of the system before exchanging ICs.

## Questions

(1) Which IC in the system would be suspect if the program allowed inputs to be read from Port A but no drive signals were available from Port B?

(2) What would be the result of a short circuit between pins 1 and 2 of the microprocessor?

(3) $\overline{CS2}$ input of the PIA is moved to the $Y_5$ output of the address decoder. What is the new base address of the PIA?

(4) Gate B fails with its output stuck at 1. What effect would this have on the system?

(5) What is the difference between an $\overline{NMI}$ and $\overline{IRQ}$ input?

(6) With the motor and switch hardware connected a fault develops so that the motor is run correctly up to max. speed but the max. speed indicator does not come on. What is the most likely fault and how can this be quickly tested?

(7) The reset switch to the CPU fails open circuit. How would this fault be indicated and what one test would verify the fault?

(8) The system and its program seem to operate correctly except for the fact that zero speed cannot be set by the input switches. What is the likely fault?

(9) The program appears to operate correctly up to the "ramp motor" section. At this point the motor starts to operate but stays at minimum speed. The max. speed indicator operates correctly. What chip or chips in the system are suspect? Detail the tests you would make to isolate the fault.

(10) Explain the effect of the following hardware faults:

(a) S5 open circuit.
(b) Bit 5 of Port A on the PIA stuck at 1.
(c) Gate A output stuck at 1.
(d) R/$\overline{W}$ line open circuit to the PIA.

# 9 The Fault Finding Process

## 9.1 Practical Tests

Many courses have included practical fault finding tests into their assessment of competences, in addition to the traditional fault finding examination paper. In these practical tests the student is given a faulty circuit or system together with suitable test equipment. He or she then has to trace the fault down to component level. This fault finding process requires the person being tested to carefully observe and record each step taken in tracing the fault. Marks are awarded for the correct sequence of steps, and in this way any element of guesswork tends to be eliminated. It is possible, of course, to pass such a test even if the component causing the fault is not identified, as long as an almost correct set of logical steps to locate the fault is clearly indicated and recorded.

As shown in all the previous chapters of this book, the skill of fault finding is basically a process that consists of a series of logical and well-defined steps. These defined actions can be written as follows and also shown as a flow chart as in Fig. 9.1.

(1) Observe and write a brief description of the initial symptoms.
(2) From these initial symptoms deduce the portion (or portions) of the system that is causing the fault. Record the reasoning behind this choice.
(3) Select a suitable measuring instrument, such as a DMM, analogue meter, CRO or logic probe and then measure at selected test points around the suspected fault region. Make a clear record of the readings.
(4) Using the recorded readings from (3) and the symptoms from (1) narrow down the fault to one component. State the reasons for the choice of the faulty component and

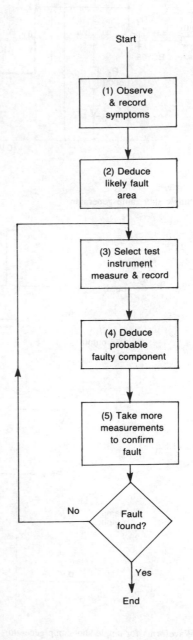

**Fig. 9.1** Flow chart showing fault finding process

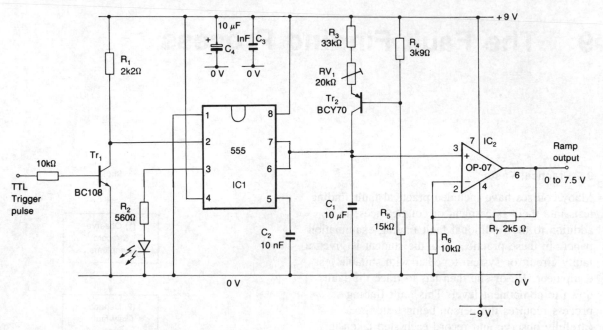

**Fig. 9.2**  Single shot ramp generator

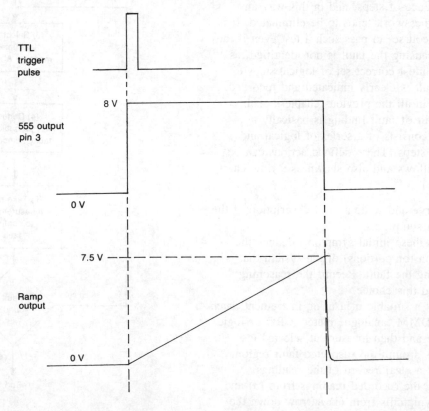

**Fig. 9.3**  Waveforms for single shot ramp generator

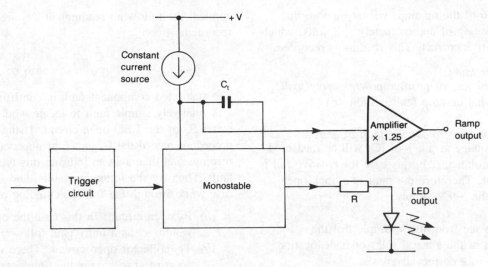

**Fig. 9.4** Block diagram of system

the probable type of failure.

(5) If necessary take more readings or different measurements to confirm the diagnosis.

The process is best illustrated by a real example. Fig. 9.2 shows a single shot ramp generator designed to be triggered by a TTL input pulse and to deliver an output of a single ramp. The ramp has an amplitude of $7 \cdot 5$ V and a duration of 2 seconds. Typical waveforms are shown in Fig. 9.3.

In the circuit, a 555 timer ($IC_1$) is wired as a monostable but the normal timing resistor is replaced by a constant current source using the p-n-p transistor $Tr_2$. The timing capacitor $C_1$ is therefore charged by a fixed value of current and the voltage across it will rise linearly with time to produce a 0 to 6 volt ramp. This ramp is amplified by a non-inverting op-amp circuit using $IC_2$. The two resistors $R_6$ and $R_7$ set the gain of the amplifier to $1 \cdot 25$ so that a 0 to $7 \cdot 5$ volt ramp output results. When triggered the circuit is arranged to give an indication of the on state duration using the main output pin of the 555, pin 3, to drive an LED via a limiting resistor. This LED should light for 2 seconds.

To simplify fault finding it is always useful to think of a circuit in block form, and a sketch of the block diagram of this circuit is shown in Fig. 9.4. This clearly indicates the main portions of the

circuit enabling faults to be isolated to one block and then traced to one component within that faulty block.

Using the procedure outlined in the previous paragraphs let us consider the following fault symptoms:

**(1) *Initial symptoms***
When the trigger pulse is applied $LED_1$ lights for 2 seconds but there is no rising ramp output. The output of the amplifier is fixed.

**(2) *Comment on the likely portion of the circuit at fault***
From the initial symptoms we can make the observation that both the monostable and the timing network appear to be functioning correctly. Therefore the fault must lie in the amplifier circuit.

**(3) *Measurement to confirm fault***
The most suitable test instruments for making measurements on this type of circuit would be a CRO and an analogue multimeter.

A trigger pulse is applied and the ramp observed, using the CRO, at the junction of $C_1$ and the collector of $Tr_2$. This will be a 0 to 6 volt ramp with a 2 second duration, showing that the 555 and the timing network are in fact correct. Using the multimeter, the next measurement will be at pin 6 of the op-amp. Let us suppose this gives a reading of approximately $-8$ volts, which is obviously incorrect. This reading is recorded.

(4) *Possible fault(s)*
A suspected loss of positive power supply to $IC_2$, or an internal op-amp fault is indicated.

(5) *Final readings to confirm fault diagnosis*
The rail voltage at pin 7 of $IC_2$ will be measured using the voltmeter. In this case the positive rail is not present. Therefore the fault is a track open circuit to the $+9$ V supply.

You can see from this example that the application of the general rules of fault location will result in a correct diagnosis.

Now imagine a fault where $Tr_2$ has failed collector-emitter short circuit. The effect would be illustrated, and the fault located, by the following sequence.

(1) *Initial symptoms* ($Tr_2$ collector-emitter short)
When the trigger pulse is applied the LED lights briefly and a short duration non-linear ramp is produced at the output of the circuit.

(2) *Comment on likely fault*
The portion of the circuit most likely to produce these symptoms is the timing circuit. Both the monostable and the amplifier appear to be functioning otherwise no ramp at all would appear.

(3) *Measurements to locate faulty component*
A CRO is selected since measurements will have to be made at the timing capacitor. The CRO is connected to the junction of $C_1$ and $Tr_2$ collector and a trigger pulse applied to the circuit. A short duration non-linear ramp lasting about 50 ms is observed. When the CRO is moved to the emitter of $Tr_2$ and another trigger pulse applied, the same non-linear ramp is observed.

(4) *Possible fault(s)*
The measurements and symptoms point to a short circuit between the collector and emitter of $Tr_2$.

(5) *Final readings to confirm fault*
Using an analogue voltmeter and with no trigger

applied, the following readings at $Tr_2$ are recorded:

| $Tr_2$ | C | B | E |
|---|---|---|---|
| Reading | 0 V | 7·1 V | 0 V |

The suspected component fault is confirmed.
A relatively simple fault to locate would be either $R_2$ or the LED open circuit. Using the same procedure record the sequence of observations, readings and diagnosis to indicate this type of fault. Then try the following faults, and in each case work through the full fault finding procedure.

(a)  $R_6$ open circuit. In this case the op-amp would act as a unity gain follower.
(b)  $Tr_2$ collector open circuit. There would be no current to charge the timing capacitor $C_1$.
(c)  $C_1$ short circuit.
(d)  $Tr_1$ base-emitter short circuit.
(e)  $R_7$ open circuit. The op-amp will be forced to act as a comparator.
(f)  Track open circuit to pin 7 of $IC_1$.

## 9.2 Exercise: DAC board (Fig. 9.5)

A digital to analogue converter is an IC that accepts an input of a digital word and produces an analogue output voltage or current proportional to the weight of the digital word at the input. In an 8-bit DAC with a reference of 2·5 volts, as used in this exercise, the input code values would produce approximate analogue voltages as follows:

| Digital value | Analogue output |
|---|---|
| 00000000 | 0 V |
| 00000001 | 10 mV |
| 00000010 | 20 mV |
| 00000011 | 30 mV |
| 00000100 | 40 mV |
| and so on | |

Each 'step' in the analogue output caused by a one-bit change (LSB) in the digital word is about 10 mV. In practical circuits the 'step', or incremental change in $V_{out}$ is given by:

$$V_{step} = \frac{V_{ref}}{2^n}$$

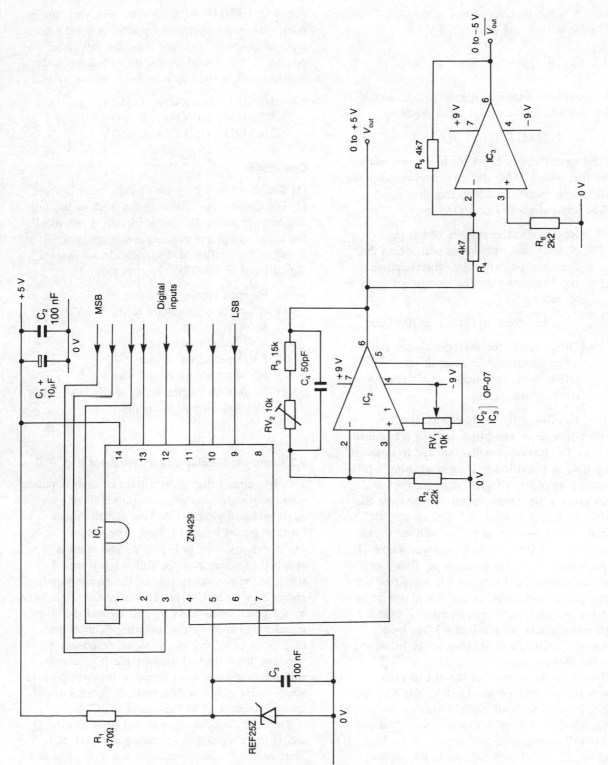

**Fig. 9.5** DAC board

where $n$ is the number of bits in the digital word. Thus for $n = 8$ and $V_{ref} = 2 \cdot 5$ V

$$V_{step} = \frac{2 \cdot 5}{256} = 9 \cdot 8 \, mV$$

The maximum analogue output voltage occurs when all bits of the digital input are high

i.e. 11111111

In this case $V_{FSO}$ is $2 \cdot 490$ V. It is worth noting here that with a DAC the maximum output voltage will be one incremental step less than $V_{ref}$.

The key parameters of a DAC are:

* Resolution: the number of bits ($n$)
* Settling time: the speed with which the analogue output settles when the digital input is changed. The change is usually in all bits

i.e. from 01111111 to 10000000

* Offset error: the analogue output value when the digital input is all zero
* Offset drift: the change of offset error with temperature.

For the ZN429 IC the resolution is 8 bits, settling time 2 $\mu$s and offset error is better than 10 mV. The analogue output voltage from the IC, on pin 4, is amplified in the circuit using a non-inverting amplifier ($IC_2$) with its gain set to 2. This gives a maximum output voltage from the circuit of nearly 4 volts. Another op-amp ($IC_3$), wired as an inverting amplifier with unity gain, is used to give a 0 to −5 volt analogue output ($\overline{V_{out}}$).

As shown in the initial example, fault location and component fault diagnosis is simplified if a circuit is not too complex and where two or more outputs are present. In this example if both analogue outputs are stuck at 0 V the fault porbably lies in the DAC chip or the reference, not the two op-amps.

To set up the outputs of the circuit to the correct levels the two potentiometers $RV_1$ and $RV_2$ are used. Initially with all digital inputs set to logic 0, $RV_1$ is adjusted to give an analogue output of 0 V ± 10 mV. Then all digital inputs are set high (1) and the $V_{out}$ level adjusted, using $RV_2$ to vary the voltage gain of the amplifier, until the analogue

output is $4 \cdot 980$ V ± 10 mV. A check can then be made on correct operation by settling the digital input to give 1/4, 1/2, and then 3/4 full scale outputs. These digital inputs are as follows and should result in the listed analogue output values.

| 1/4 | (64) | 01000000 | $1 \cdot 250$ V |
| 1/2 | (128) | 10000000 | $2 \cdot 500$ V |
| 3/4 | (192) | 11000000 | $3 \cdot 750$ V |

**Questions:**

(1) Sketch a block diagram of the circuit board.
(2) For each of the following faults show the full diagnosis process. This must include a record of the initial symptoms and required measurements to arrive at the confirmed diagnosis. In all cases a digital input of 10000000 is assumed.

(a) $R_1$ open circuit
(b) $IC_1$ track open circuit to the +5 V rail
(c) $R_2$ open circuit
(d) $RV_2$ open circuit track
(e) $RV_2$ open circuit wiper
(f) $IC_3$ short circuit pins 2 and 3
(g) Bit 6 of the digital input to the DAC stuck high (an internal IC fault)

**9.3 Exercise: Bipolar pulse generator** (Fig. 9.6)

This is a circuit that gives a train of bipolar pulses with the positive and negative pulses having an equal width of typically $6 \cdot 5$ ms and the space between pulses being $13 \cdot 5$ ms. The output waveform is shown in Fig. 9.7. The input is assumed to come from the full wave rectified signal at the secondary side of the power supply and is 20 V pk at 100 Hz. The amplitude is reduced by the potential dividers $R_1$ and $R_2$ and the waveform shaped by the inverting Schmitt gate $IC_{1a}$. A bistable divides the signal frequency by two, and the output of the bistable $IC_{2a}$ is gated with the original shaped signal to produce positive-going 50 Hz pulses with a mark to space ratio of approximately 1 : 3 at the output of $IC_{1c}$.

A second divide-by-two circuit produces the $Q$ and $\overline{Q}$ drive signals to an analogue switch $IC_3$. Thus switch X closes while switch Y is open and vice versa. When switch X closes with $\overline{Q}$ high, the

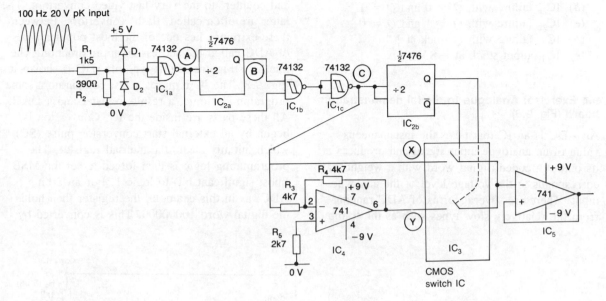

**Fig. 9.6**   Bipolar pulse generator

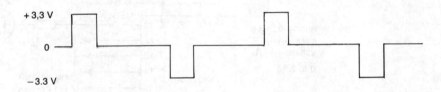

**Fig. 9.7**   Output waveform (50 Hz)

positive-going 50 Hz pulse is connected to the output of the circuit via the buffer $IC_5$, but when $\bar{Q}$ goes low switch Y closes and the negative-going pulse from the unity gain inverting amplifier $IC_4$ is sent to the output via the buffer. In this way a composite bipolar output waveform is generated. With this type of system a total loss of output could be caused by a fault in any component part.

## Questions

(1) Sketch the waveforms you would expect to measure using a CRO at the following test points.

   (a)  Output $IC_{1b}$
   (b)  Output $IC_4$
   (c)  $Q$ and $\bar{Q}$ outputs of $IC_{2b}$

(2) The circuit fails to deliver any output pulses.

You decide to use the half-split method to locate the fault.

   (a)  Which portion of the circuit do you measure first?
   (b)  If the fault was $D_1$ short circuit, how many measurements are required to locate the fault?
   (c)  Assuming that the first half of the circuit board is functioning correctly and that a check at $IC_{2b}$ output shows that this bistable is also correct, state the likely faulty component(s).

(3) For all the component failures listed below describe the fault symptoms and write the sequence of tests needed to locate the faulty component.

   (a)  $IC_3$ switch X fails open circuit

(b) $IC_{2b}$ failure with $Q = 0$ and $\bar{Q} = 1$

(c) $IC_{2b}$ failure with $Q = 1$ and $\bar{Q} = 0$

(d) $IC_{2a}$ failure with $Q$ stuck at 1

(e) $IC_4$ output stuck at $-8\,V$

### 9.4 Exercise: Analogue to digital demonstration board (Fig. 9.8)

An ADC is an IC that takes the instantaneous value of an analogue input signal and produces at its output a coded digital word with a weight that corresponds to the voltage level of the analogue input. There are several forms of ADC ranging from the relatively slow types such as the 'ramp and counter' to the very fast video convertors. The latter are often called 'flash' convertors. Between these extremes lies one of the most popular types of ADC called the successive approximation. The internal structure of this type of ADC is shown in Fig. 9.8. The IC requires a fast comparator, some programming logic, a results register and a DAC. All these parts are inside the IC. Conversion is begun by an external start conversion pulse (SC) which initially clears the internal register. The programming logic is then forced to set the MSB (most significant bit) to logic 1. For an 8-bit ADC, as in this example, the register then holds the digital word 10000000. This is converted by

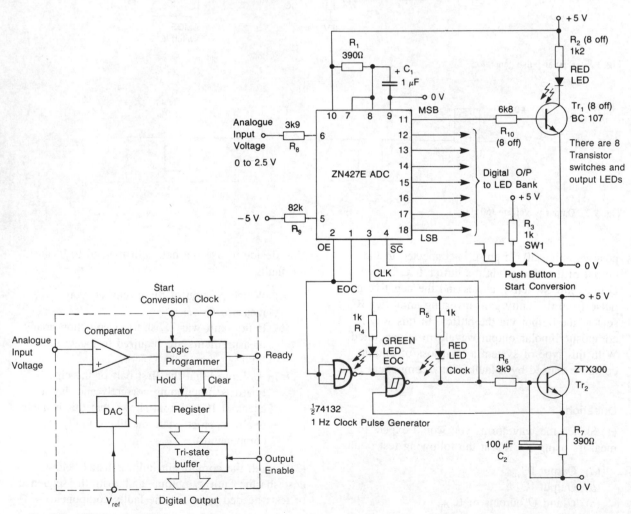

**Fig. 9.8** ADC demonstration board

the DAC and the output is compared with the analogue input. If the output by the DAC is larger than $V_{in}$, the logic 1 is removed from the MSB and placed in the next most significant bit; the register now holds 01000000. This in turn is converted by the DAC and the value compared with the analogue input. Suppose it is less than $V_{in}$: the logic 1 in that position is retained and the next most significant bit is also set to 1 (register holds 01100000) and used for comparison.

This process continues until all bits have been tried and a point of balance is reached; that is, when $V_{in}$ is just greater than the DAC output voltage.

The ZN427E 8-bit successive approximation ADC is a good example. This chip has all the necessary logic, a DAC, a $2 \cdot 5$ V precision reference and a fast comparator. Only a few external components are required to create an accurate, high-speed ADC. Conversion time can be as fast as $10\,\mu s$. In addition, tri-state output buffers are included to allow direct connection to the data bus of a system.

Brief specification details on the ZN427E are:

| | |
|---|---|
| Resolution | $= 8$ bits |
| Linearity error | $= \pm\ 0 \cdot 5$ LSB max |
| Internal voltage reference | $= 2 \cdot 560$ V typ |
| | $\left(\begin{array}{l} R_{ref} = 390\ \Omega \\ C_{ref} = 4\ \mu 7\ F \end{array}\right)$ |
| | $2 \cdot 475$ V min |
| | $2 \cdot 625$ V max |
| Slope resistance of $V_{ref}$ | $= 2\ \Omega$ max |
| $V_{ref}$ temperature coefficient | $= 50$ ppm/$^\circ$C |
| Maximum clock frequency | $= 900$ kHz |
| Start conversion pulsewidth | $= 250$ ns |
| Conversion time | $= 10\ \mu s$ (clock set to 900 kHz) |

In this circuit the ADC is wired up as a demonstration unit with the external clock slowed down and the digital output value displayed on 8 LEDs. The clock circuit is formed around two of the gates in a quad 2 input Schmitt IC (74132) with $Tr_1$ used to buffer the timing components. A description of this oscillator is given in exercise 8.8.

The start conversion pulse ($\overline{SC}$) is given via $SW_1$ which pulls pin 4 of the ZN427E low. The end of conversion (EOC) output from the ADC chip then goes low which enables the clock to run. At the same time the output enable pin of the IC, also connected to the EOC output, goes low and the output buffers inside the ADC are put into tri-state mode. Thus all the LEDs indicating the digital output value are switched off. The clock operates giving the nine pulses necessary to drive the ADC through its successive approximation cycle. These pulses are indicated by another LED. At the end of the cycle, when the ADC has performed the conversion, the IC sends the EOC output line high and the green LED lights. The clock is turned off and the OE input enabled so that the digital output value is displayed on the 8 red LEDs. If the analogue input is changed and the conversion operation repeated the change in digital output will be indicated. Note that the IC, and the circuit, is capable of fast operation. In this example the speed has been deliberately set to a very slow value so that the operation of the system can be examined.

## Questions

(1) What value of analogue input voltage should result in the following digital output words?

    (a) 10000000
    (b) 00101001
    (c) 11111110

(2) A fault exists where test inputs of $1 \cdot 25$ V and 0 V give the digital outputs 10010000 and 00010000 respectively. State which portion of the circuit is at fault and list the possible component faults.

(3) State the likely symptoms and the full fault finding procedure for the following component faults

    (a) $C_2$ open circuit
    (b) $R_8$ open circuit
    (c) Open circuit track to pin 2 (OE) of the ADC
    (d) Pin 13 of the ADC open circuit track to the driver transistor
    (e) Pin 16 of the ADC stuck at 1
    (f) $R_4$ open circuit

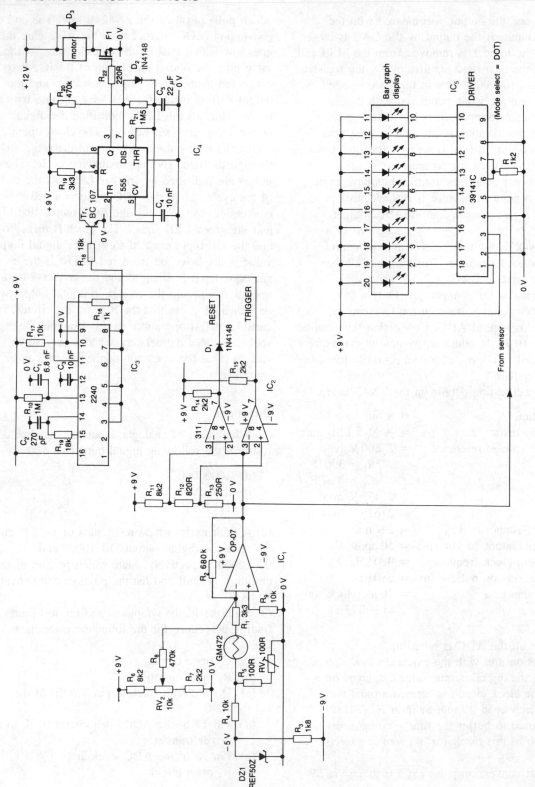

**Fig. 9.9** Process controller

## 9.5 Exercise: Process controller (Fig. 9.9)

This controller is designed for an industrial plant where an agitation process, driven by a geared 12 V 2 A d.c. motor, is operated 'on' and 'off' for a 25 minute period. This occurs while the temperature of the mix being agitated falls below 12 °C. The process is halted if the temperature of the mix rises above 18 °C. A thermistor type GM472 is used as the temperature sensor and this is arranged in a circuit to give a linear output voltage to drive a simple 10 segment bar graph display. The 'on' and 'off' periods for agitation are nominally 10 seconds and 25 seconds.

The system, shown in block diagram form in Fig. 9.10, consists of the temperature sensing circuit, two comparators, a long duration timer, an oscillator with its output driving the motor via a power switch, and a display driver and 10 segment display for temperature indication.

Briefly, the specification is:

| | |
|---|---|
| Power supply: | ± 12 V |
| Amplified output: | Linear 0 V to 1·25 V to drive the display IC and the comparators |
| Accuracy of output: | 0·25 V ± 0·1 V at 12 °C input 1·00 V ± 0·125 V at 18 °C input |
| Comparator trip levels: | 0·25 V for trigger 1·00 V for reset |
| Main timer: | 25 minutes ± 2·5 minutes |
| Astable: | 10 sec on ± 2 sec 25 sec off ± 2 sec |
| Power switch output: | 12 V 2 A |
| Display: | Dot mode, from 11 °C to 20 °C inclusive |

The thermistor has been specified because of its good sensitivity. To enable its changes in resistance to be converted to linear voltage output it is placed in the input lead of an inverting op-amp circuit in series with a 'padding' resistor. The gain of the inverting amplifier is given by:

$$A_{vcl} = \frac{-R_F}{(R_T + R_1)}$$

where $R_T$ is the thermistor's resistance
$R_1$ is the padding resistance
$R_F$ is the feedback resistor

Then:

$$A_{vcl} \text{ is proportional to } \frac{-1}{R_T + R_1}$$

In this way the non-linearity of the sensor's changes in resistance with temperature will be made to give an almost linear output voltage.

The input to the amplifier is set to a stable value using a 5 V band-gap reference REF50Z which has a 40 ppm temperature coefficient. With its current set to 2 mA (the maximum is 5 mA):

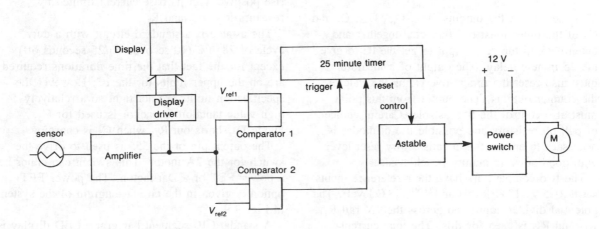

**Fig. 9.10** Block diagram of system

$$R_3 = \frac{(V_S^- - V_Z)}{I_Z}$$

Therefore $R_3 = 2\,k\Omega$; a $1k8\,\Omega$ is specified

The $-5\,V$ is reduced to about $-75\,mV$ by $R_4$, $R_5$ and $RV_1$. $RV_1$ is used to set the overall output level of the circuit.

When the temperature being sensed is $10\,°C$ the output of the amplifier is arranged to be zero volts. An offset adjustment circuit using a resistor connected to a portion of the positive rail does this. To ensure that all tolerances are covered this is made a preset using the potential divider chain $R_6$, $RV_2$ and $R_7$. With $RV_2$ in mid track the positive voltage at the input end of $R_8$ will be $3\,V$.

The 2240, connected across the $+9\,V$ supply, has the monostable period set by $R_{10}$ and $C_1$ with the overall time delay given by:

$$T = NR_{10}C_1$$

In this case $T = 1500$ seconds. To arrive at the maximum values for the RC time constant imagine $N$ is at 255 (all outputs on the 2240 used).
Then:

$$R_{10}C_1 = 1500/255 = 5\cdot88 \text{ seconds}$$

To allow for tolerance effects the RC value is made higher than this, at $6\cdot8$ seconds by using $R_{10}$ as $1\,M\Omega$ and $C_1$ as $6\cdot8\,\mu F$. Then

$$N = 1500/6\cdot8 = 220$$

This means that the outputs $O_{128}$, $O_{64}$, $O_{16}$, $O_8$ and $O_4$ of the timer must be connected together and taken back to the reset input to get the IC to give the 25 minute delay. The output of comparator 2 must also reset the timer, overriding the state of the counter outputs. The output from this point must also control the 555 astable. Careful choice of pull-up resistors and possible use of diodes is necessary to ensure that a satisfactory reset level with good noise immunity is achieved.

The two comparators have their reference inputs set to $0\cdot5\,V$ (12 C level) and $1\,V$ (18 C level). The potential divider connected across the 9 V rail $R_{11}$, $R_{12}$ and $R_{13}$ is used for this. The total current through the chain is set to about $1\,mA$ so that the values required are $8\,k\Omega$, $750\,\Omega$, and $250\,\Omega$ respectively. The two comparators, which have open collector outputs, use pull-up resistors to the $+9\,V$ rail.

The output from comparator 1, the trigger level, is connected directly to pin 11 of the 2240. Comparator 2's output, which will rise high if the 18 C trip value is exceeded, is connected via signal diode $D_1$ to the junction of $R_{16}$ and $R_{17}$, the reset pin of the 2240. The values of the components have to be chosen to ensure that a satisfactory reset is possible both from the counter outputs of the IC or from the comparator. When the comparator output is low $D_1$ is reversed so that the reset pin is at a voltage given by:

$$V_R \approx \frac{V_{CC}R_{16}}{R_{17} + R_{16}}$$

Therefore $V_R \approx 0\cdot9\,V$

This gives a noise margin on the reset of about $500\,mV$ since the reset threshold is specified at $1\cdot4\,V$.

The switching transistor is used to control the 555 astable and is connected from pin 4 (the 555 reset) to ground. A pull-up of 3k3 is used from pin 4 to the $+9\,V$ rail.

$Tr_1$ base is connected back to the command output lines of the 2240 via $R_{18}$. While the timer is in the monostable mode $Tr_1$ will therefore be off, allowing the 555 oscillator to run. At the end of the timing period, or after a reset, $Tr_1$ base will rise positive with its base current limited by resistors $R_{17}$, $R_{16}$ and $R_{18}$.

The astable is a standard circuit with a duty cycle of $28\cdot6\%$ (10 seconds on/25 seconds off) except for the fact that the time durations required are on the upper limits for the IC. However the specification limits are not tight so a relatively high value tantalum capacitor is used for $C_5$. Diode $D_2$ shorts out $R_{21}$ when $C_5$ is charging.

The output pin of the 555 is used to drive the switch for the 2A motor. This switch can either be a power FET or a Darlington. The power FET option is given in the circuit diagram of the system in Fig. 9.9.

A standard 10 segment bar graph LED display is used for this part of the circuit and this is driven by a 3914 IC. This chip contains all the necessary comparators, a precision divider chain, a stable $1 \cdot 12$ V reference and LED switches. With pins 6 and 7 connected together the input range is from 0 to $1 \cdot 25$ volts. In other words, every 125 mV change on the input line will cause the next comparator in the chain to operate. One resistor, $R_{22}$, fixes the current through each LED.

$$\text{LED current} \approx \frac{10 V_{ref}}{R_{22}}$$

with $R_{22} = 1k2 \ \Omega$ each LED will have an 'on' current of 10 mA.

Pin 9 of the IC determines the mode of display. For 'dot' mode this pin is left open circuit.

## Questions

(1) Show how a simple modification can be made to the circuit to provide a manual reset.

(2) Describe the symptoms caused by an open circuit thermistor. Show how this component failure could be located.

(3) List the symptoms and the required fault location sequence for the following component faults.

(a)  $RV_1$ open circuit
(b)  $F_1$ short circuit drain to source
(c)  $Tr_1$ base–emitter short circuit
(d)  $D_1$ open circuit
(e)  $Tr_1$ collector–emitter short circuit
(f)  Open circuit track from pin 13 of $IC_5$ to the bar graph display
(g)  $R_{22}$ open circuit

# Answers to Exercises

## Chapter 2

EXERCISE 1 (page 23)
Estimation of meter indications in circuits given in Fig. 2.14. Your answer should be within ±10% of that given below.

| Fig. 2.14A | TP | 1 | 2 | |
|---|---|---|---|---|
| | MR | +3 | +2·3 | |

| Fig. 2.14B | TP | 1 | 2 | 3 |
|---|---|---|---|---|
| | MR | −3 | −7·8 | −2·3 |

| Fig. 2.14C | TP | 1 | 2 | |
|---|---|---|---|---|
| | MR | −0·3 | −1·2 | |

A base current of 25 μA will flow through $R_1$ causing the base voltage to be approximately −0·7 V. Connecting a meter to the base will lower this voltage.

| Fig. 2.14D | TP | 1 | |
|---|---|---|---|
| | MR | +18 | |

EXERCISE 2 (page 23)

| Normal voltages | TP | 1 | 2 | 3 |
|---|---|---|---|---|
| | MR | 1·5 | 4·5 | 0·8 |

Fault
  A  Base emitter short circuit
  B  $R_4$ open circuit
  C  Collector open circuit
  D  Collector emitter short
  E  $C_3$ open circuit giving negative feedback
  F  $R_1$ open circuit.

EXERCISE 3 (page 23)

| Normal voltages | TP | 1 | 2 | 3 |
|---|---|---|---|---|
| | MR | 2·3 | +3 | +7 |

Fault
  A  $R_2$ open *or* $C_3$ short circuit
  B  Base emitter open circuit
  C  Collector emitter short
  D  $R_4$ open circuit
  E  Collector base short
  F  $R_3$ open circuit.

EXERCISE 4 (page 24)
(a) $C_1$ open circuit. The RF should be filtered out.
(b) $C_2$ open circuit.
(c) $C_1$ open circuit. $R_1 C_1$ should form an integrator of 1 ms time constant.
(d) Diode open circuit. The diode should clip the positive spike.
(e) $L_2$ open circuit. Signal present at collector shows that the colpitts oscillator is working correctly.

## Chapter 3

EXERCISE 3.7 (page 31)
Fault
  A  DZ open circuit. Voltage at TP3 has risen. No current flowing through $R_2$.
  B  $C_1$ open circuit. D.C. level at TP1 lower
  C  $R_3$ open or high *or* $Tr_1$ base emitter open circuit
  D  $R_1$ open circuit *not* $C_2$ short, because TP1 is high
  E  DZ short circuit
  F  $Tr_2$ base emitter open circuit
  G  $R_4$ open or high in value
  H  $Tr_1$ collector emitter short circuit.

EXERCISE 3.8 (page 33)
  A  $R_1$ open circuit or DZ short circuit
  B  $R_6$ or $RV_2$ open circuit. Possibly $C_2$ short
  C  $Tr_3$ base emitter open circuit (or $Tr_4$)
  D  $R_3$ open circuit
  E  $RV_1$ or $Tr_5$ base open circuit or $R_8$ open circuit
  F  $Tr_2$ base emitter open circuit
  G  $Tr_4$ short collector to base
  H  Transformer secondary or primary open circuit
  I  Transformer primary short circuit to screen.

EXERCISE 3.9 (page 35)

(1) A  $R_{17}$ open circuit      B  $L_1$ open circuit
    C  Resistance checks are normal. Fault could be open circuit component in $Tr_7$, $Tr_8$, $Tr_9$ circuit, i.e. $R_{13}$ open circuit.

(3) (a) $C_1$ s/c. Fuse blown. TP1 resistance to O V, zero $\Omega$.

(b) No output, although oscillator will start. No switching signal at $Tr_8$ collector.

(c) $Tr_2$ ON and $Tr_1$ OFF. Therefore no forward bias for $Tr_8$. $Tr_9$ will be non-conducting and the output zero.

(d) $Tr_9$ cannot conduct, so output remains zero. However the oscillator will run when the start switch is pressed.

(e) Circuit will operate and output will stabilize at 20 V only when the start switch is pressed.

(4) Resistance readings normal. Fault is primary short to screen

## EXERCISE 3.10 (page 37)
Fault

A $L_1$ open circuit

B $RV_1$ open circuit

C $R_1$ open circuit

D Internal switch transistor open circuit (Pin 16)

E Break in track pin 8 to pin 9.

## EXERCISE 3.11 (page 38)
Fault

A $R_1$ open circuit

B $RV_1$ track open circuit

C $C_2$ short circuit

D $D_2$ short circuit

E Adjust pin open circuit

## EXERCISE 3.12 (page 39)

(1) To prevent noise or spikes falsely triggering the thyristor

(2) $V_T$ max $\approx 15\cdot65$ V

$V_T$ min $\approx 14\cdot35$ V

(3) Nearly zero, since the zener $DZ_1$ should be non-conducting

(4) (a) The trip voltage will be low at approximately 13 V.

(b) The overvoltage trip protection circuit will not function since the thyristor gate voltage will always remain at zero volts irrespective of $V_O$.

(c) The fuse will blow at switch on.

## EXERCISE 3.13 (page 39)

(1) | Fault | TP | 1 | 2 | 3 | 4 | 5 |
|---|---|---|---|---|---|---|
| (a) | V | 0 | 0 | 0 | 0 | 0 |
| (b) | V | 7·18 | 7·18 | 7·46 | 8·32 | 7·18 |
| (c) | V | 7·00 | 5·93 | 0·7 | 0·75 | 0·002 |
| (d) | V | 0 | 0 | 0 | 0 | 0 |
| (e) | V | 7·08 | 5·94 | 5·94 | 6·98 | 5·94 |

(2) | | TP | 1 | 2 | 3 | 4 | 5 |
|---|---|---|---|---|---|---|
| (a) | V | 7·08 | 5·93 | 3·21 | 3·78 | 3·21 |

Here the current limit reverts to the simple overcurrent type with $I_{SC}$ set to about 55 mA

| | | | | | | |
|---|---|---|---|---|---|---|
| (b) | V | 7·08 | 5·93 | 0 | 6·98 | 5·93 |

No current limit is in operation

## Chapter 4

## EXERCISE 4.7 (page 53)
Fault

A $C_3$ short circuit. This causes $Tr_2$ to conduct heavily and cuts off $Tr_1$

B $R_5$ open circuit. Causes very high gain and change in d.c. bias levels

C $Tr_1$ collector emitter short

D $C_2$ short circuit

E $Tr_2$ collector open circuit

F $C_3$ open circuit. Introduces negative feedback

G $R_1$ open circuit.

## EXERCISE 4.8 (page 55)
Fault

A $R_4$ open circuit. Note that when the meter is connected in circuit it presents a current path for TP3, 4 and 5.

B $C_1$ short circuit. TP3 and 4 are at the same voltage.

C $C_1$ open circuit, which introduces negative feedback to reduce gain.

D Drain of FET open circuit. $Tr_2$ cannot conduct.

E $Tr_2$ base emitter open circuit.

## EXERCISE 4.9 (page 57)
Fault

A Zener diode $DZ_1$ open circuit

B $RV_1$ wiper open circuit

C $R_7$ open circuit. No negative feedback
D Zener $DZ_1$ short circuit
E $Tr_1$ base emitter open circuit.

EXERCISE 4.10 (page 59)
(1) Fault
   A $R_8$ open circuit
   B $R_4$ open circuit
   C $R_3$ open circuit or $Tr_1$ base emitter open.
   D $D_1$ or $D_2$ short circuit
   E $Tr_3$ base emitter short circuit
   F $R_2$ open circuit

*Chapter 5*
EXERCISE 5.9 (page 68)
(1) $C_1$ or $C_6$ open circuit
(2) $R_3$, $RV_{1B}$, or $C_8$ open circuit. $R_3$ or $RV_{1B}$ open would prevent positive feedback necessary for oscillations. $C_8$ open would introduce more negative feedback.
(3) (*a*) High gain giving grossly distorted output.
    (*b*) No output on switch position 1.
(4) $R_1$ open circuit
(5) $C_2$ short circuit
(6)

| TP | 1 | 2 | 3 | 4 | 5 |
|----|-----|-----|-----|-----|-----|
| MR | +1·9 | +0·8 | +0·8 | 0 | +0·1 |

(7) $R_7$ open circuit.

EXERCISE 5.10 (page 69)
(1) $R_1$ open circuit or $C_1$ short circuit.
(2) $C_2$ open circuit.
(3) $C_1$ open circuit or secondary winding short circuit.
(4)

| TP | 1 | 2 | 3 | 4 |
|----|-----|-----|-----|-----|
| MR | 0·7 | 0·7 | 12 | 12 |

(5) The time constant of $R_2C_2$ will increase. This will not change frequency, but will reduce sawtooth amplitude.
(6) Transformer secondary open circuit.

EXERCISE 5.11 (page 71)
(1) $D_1$ short circuit. This increases current through $Tr_3$ and therefore almost doubles the frequency.
(2) Emitter follower action would cease. Linearity of output would be poor.
(3) A $R_7$ open circuit
   B $D_1$ or $R_9$ open circuit
   C $Tr_2$ base emitter open circuit
   D $R_2$ open circuit
   E $Tr_3$ collector open circuit.
(4) $R_6$ open circuit. As soon as voltage across $C_1$ rises more than 0·8 V, $Tr_1$ and $Tr_2$ switch

EXERCISE 5.12 (page 73)
(1) Modify $Tr_1$ and $Tr_2$ circuit into a low speed astable multivibrator.
(2) A.C. feedback to ensure rapid switch off at the end of the ramp time.
(3) A $D_2$ open circuit no feedback to switch bistable. Therefore output remains high.
   B $R_5$ open circuit
   C $C_4$ short circuit
   D $Tr_2$ base collector short
   E $D_1$ open circuit. No trigger pulse reaching $Tr_1$.
   F $Tr_3$ base emitter short
   G $Tr_3$ base collector short
   H $R_{10}$ open.

EXERCISE 5.13 (page 74)
(1) Output A normal, no pulses from B.
(2) $C_3$ open circuit.
(3) $Tr_1$ would no longer act as a transistor. $Tr_2$ collector voltage would fall. So output A would be at zero volts, and output B would remain at +8 V.
(4) $Tr_2$ base emitter short circuit or $R_3$ open circuit.
(5) A $Tr_3$ emitter open circuit or $R_5$ open or $C_3$ short circuit
   B $Tr_4$ emitter to $B_1$ short
   C $Tr_4$ emitter open circuit
   D $Tr_5$ base emitter short circuit
   E $R_6$ open circuit.

EXERCISE 5.14 (page 75)
(2) (*a*) Operation of the switch will not trigger the monostable. The output will remain low and the voltage at pins 6 and 7 will be zero.

(b) The circuit will operate as a monostable but the pulse width will be too brief to be indicated as it depends upon the value of stray capacitance from pins 6 and 7 to ground. Locate the fault by bridging $C_1$ with a good 10 $\mu$F capacitor.

(c) The timing period, after operation of the switch, will be exceptionally long. $C_1$ will try to charge via any leakage resistance towards 9 V. The output will therefore remain high. Reset will be operational.

(d) No triggering possible, therefore the output will remain low.

(e) The trigger input will be permanently low forcing the output to be in a high state.

## EXERCISE 5.15 (page 76)

(1) $f = 21 \cdot 258$ kHz.

(2) Check output of 555 astable.

(3) Using a logic pulser inject signals at pin 1 of the 4024B.

(4) Bridge $C_1$ with a relatively high value capacitor.

(5) An open circuit in the R-2R ladder, most probably the 10 k$\Omega$ resistor nearest the output.

## Chapter 6

## EXERCISE 6.8 (page 86)

(1) $D_1$ short circuit. Measurement with ohmmeter.

(2) $D_2$ open or $R_5$ open. Measure d.c. level at $R_4R_5$ junction.

(4) Check presence of signal at junction $R_2$, $D_1$ and $D_2$.

(5) $D_1$, $C_1$ or $R_2$ open circuit. The diode would have the higher failure rate.

(6) $D_2$ short circuit.

(7) $R_4$ open circuit.

## EXERCISE 6.9 (page 87)

(1) $R_5$, $C_1$.

(2) Speed-up capacitor.

(3)

| TP | 1 | 2 | 3 | 4 | 5 |
|---|---|---|---|---|---|
| MR | just +ve | +7 | +0.7 | +0.1 | +6.2 |

(4) Momentarily short base to emitter. Collector voltage should rise to +7 V.

(5) A   $R_1$ open circuit

B   $C_1$ open circuit (possibly $C_2$ or $D_3$ open)

C   $R_4$ open circuit or $Tr_2$ b/e short

D   $DZ_1$ short circuit

E   $Tr_3$ base emitter short or $R_5$ open.

(6) (a) Bias voltages normal. No output pulse since input cannot reach $Tr_1$ base.

(b) Same conditions as above.

(c) TP4 rises to +7 V. This forward biases $Tr_2$.

| TP | 1 | 2 | 3 | 4 | 5 |
|---|---|---|---|---|---|
| MR | 0.7 | 0.13 | 0.7 | 7 | 6.2 |

(7) $D_2$ open circuit.

(8) Take signal from $Tr_2$ collector via a capacitor.

## EXERCISE 6.10 (page 88)

(1)

| TP | 1 | 2 | 3 | 4 | 5 |
|---|---|---|---|---|---|
| MR | 1.5 | 0.95 | 0.2 | 0.9 | 12 |

(2)

| TP | 1 | 2 | 3 | 4 | 5 |
|---|---|---|---|---|---|
| MR | 0 | 8.3 | 1.8 | 1.1 | 6.7 |

Output will not be perfect square wave. Mark-to-space ratio will be low.

(3) A   $Tr_1$ base collector open

B   $Tr_2$ base emitter open or $C_1$ open

C   $R_4$ open

D   $Tr_2$ base emitter short

E   $Tr_1$ base collector short

F   $R_5$ open

## EXERCISE 6.11 (page 88)

(1) Temporarily connect a 10 k$\Omega$ resistor from +5 V to transistor base. Collector voltage should fall to nearly zero.

(2) Very low noise margin, particularly with switch in '0' position

(3) A   $L_1$, $R_1$ or $R_2$ open circuit. Possibly $C_2$ short.

B   $R_4$ open or collector emitter short.

C   $Tr_1$ Base emitter short or $D_1$ short

D   $Tr_1$ base emitter open.

(4) 0 V   −0.6 V   +5 V

(5) 30 V   +0.75 V   +0.75 V   Switch '1'

0 V   +0.75 V   +0.75 V   Switch '0'

EXERCISE 6.12 (page 90)

(1) Electronic voltmeter with very high input impedance.

(2) $R_1$ open. $C_1$ short.

(3) $Tr_4$ clamped to +10 V. Output always 0 V.

(4) X   $DZ_1$ open

Y   $Tr_2$ collector base short

Z   $Tr_1$ gate to source short

(5) $R_4$ open.

*Chapter 7*

EXERCISE 7.6 (page 98)

(1)
| Lamp TP | 1 | 2 | 3 | 4 | 5 |
|---|---|---|---|---|---|
| ON | 3 | 3·6 | 3·7 | 0 | 12 |
| OFF | 5·8 | 4·4 | 3·7 | 0·7 | 0·9 |

(2) Alarm operating continuously. Cannot be reset.

(3) A   $SCR_1$ open circuit gate

B   $R_1$ or $RV_1$ open circuit

C   $SCR_1$ anode to cathode short

D   $R_5$ open or high

E   $Tr_1$ collector base short

F   $Tr_2$ collector base short

(4) $R_6$ open or failure of the reset switch. Bridge $R_6$ with 1k resistor.

EXERCISE 7.7 (page 99)

(1) RF filter.

(2) 1·4 kW

(3) Lamp very dim. *No* control.

(4) Triac short MT1 to MT2.

(5) $R_2$ or $R_3$ open ⎫ Bridge suspect component with

(6) $C_2$ open          ⎭ another of same value.

(7) Lamp unlit. With gate open, a larger a.c. voltage will be indicated at TP3 than when gate to cathode is short.

(8) $L_1$ open.

EXERCISE 7.8 (page 101)

(1) Electronic voltmeter or chart recorder.

(2) Use formula $v_c = V(1 - e^{-t/CR})$

Then $t \simeq 0.88\,CR$

With $RV_1$ min. $t \simeq 0.88$ sec.

With $RV_1$ max. $t \simeq 18$ sec.

(3) Timing circuit formed by $Tr_3$ and $UJT_3$. $RV_3$ open, $C_3$ short, $Tr_3$ or $UJT_3$ open.

(4) Use diodes to sense state of $SCR_2$ and $SCR_3$ anode potentials and to gate signal to $SCR_1$.

(5) $C_5$ short.

(6) $D_1$ open.

(7) $SCR_2$ anode to cathode short *or* UJT2 fault such as $B_1$ to $B_2$ short. Measure d.c. voltage at SCR2 anode. If zero, $SCR_2$ at fault.

EXERCISE 7.9 (page 102)

(1) Approx 12 Hz max, 1·5 Hz min.

(2) $C_3$ short.

(5) A   $C_4$ short, $RV_1$ or $R_9$ open

B   $D_1$ or $C_1$ open

C   $R_1$ open

D   UJT $B_2$ to $B_1$ short

E   $C_2$ or $D_2$ open circuit.

F   $R_7$ open

(6) Replace start switch with high value resistor (68 kΩ) bridged by a 0·1 μF capacitor.

(7) Fit low value resistors or thermistors in series with lamps.

EXERCISE 7.10 (page 103)

(2) $SCR_1$ short anode to cathode.

(3) $SCR_1$ open circuit anode or gate. Possibly $R_1$, $RV_1$, $D_2$, etc. open. Apply gate signal direct to SCR to localize fault.

EXERCISE 7.11 (page 104)

(1) $I_F = \dfrac{V_S - (V_F + V_{OL})}{R_1}$

Here $V_{OL}$ = Low state logic output $\approx 0.4$ V

$\therefore I_F = 13.3$ mA

(2) (a) 100 millisec

(b) 400 millisec

(c) 2 Hz

(d) 1:4

(3) (i) Check LED operation.

(ii) Test mains supply voltage is present either side of fuse.

(iii) Apply short across main triac and check that load current flows.

(iv) Short pins 6 and 4 of the OPI 3041 and check that load current flows.

(4) (a) LED off pin 1 at 0 V, pin 2 just above 0 V.

(b) LED off pin 1 at 5 V, pin 2 just above 0 V.

(c) LED on continuously, therefore full power applied to load irrespective of logic input state.

(5) The LED must be in an off state, therefore the fault lies in the "mains" portion of the circuit. A short circuit main triac is the most probable fault. To check, short gate to $MT_1$ and check that load is off.

## Chapter 8

### EXERCISE 8.5 (page 111)

(1)

| TP | 1 | 2 | 3 | 4 | 5 | 6 |
|----|----|----|----|----|----|----|
| MR | +2 | +7·5 | +11 | +1·9 | +1·2 | +0·6 |

(2) 741 output open circuit.

(3) A   741 output short to +12 V
    B   $Tr_1$ short base to collector
    C   $RV_1$ wiper open circuit

### EXERCISE 8.6 (page 112)

(1) $P_{1b}$

(2) Decoupling capacitors. Located near IC.

(3) $P_{1a}$ or $P_2$, $P_3$.

(4) Measured against a very stable frequency standard for accuracy. For drift, use a frequency counter of high resolution.

(5) (a) No 100 Hz output, all other outputs correct.
    (b) No outputs from any section.
    (c) Only the 1 MHz output present.

(6) Inject a suitable (TTL-compatible) signal at $P_2$ input and use a CRO to check for correct outputs from $P_2$ to $P_5$.

### EXERCISE 8.7 (page 113)

(1) Either output open circuit or short pin 6 to pin 4.

(2)

| Pin no. | 2 | 3 | 7 | 6 | Output |
|---------|------|------|-------|------|--------|
| A | 0 | +5·7 | 0 | 0 | 0 |
| B | 0 | +5·7 | +16·2 | +11 | +10·4 |
| C | +5·7 | +5·7 | +16·2 | +6·4 | +5·7 |
| D | +0·2 | 0 | +16·2 | +1 | +0·4 |

### EXERCISE 8.8 (page 115)

(2) (a) $R_3$ open circuit
    (b) $C_2$ open circuit
    (c) $R_1$ open circuit; output of Schmitt A stuck at 1; $C_1$ short circuit; or $Tr_1$ base emitter open circuit.
    (d) $C_1$ open circuit.

(3) (a) Point (x) held at logic 0; thus, no output.
    (b) Same symptoms as (a).

(c) Input to Schmitt B will assume logic 1 state giving continuous 20 kHz pulses from output.
    (d) No output from (y).
    (e) 1 kHz pulses from (x) only.

(4) Increase values of $C_1$ and $C_2$ to give, say, 2 Hz from (x) and 1 kHz from (y). Use a transistor buffer (BFY50 or similar) to drive a small speaker from (y).

### EXERCISE 8.9 (page 116)

(1) Approx. 650 Hz.

(2) $C_1$ open circuit.

(3) (a) No oscillations. $IC_1$ output drift up to about +8 V (taking about 2 secs from switch-on) and $IC_2$ output is fixed at +8 V.
    (b) $IC_1$ output held at zero volts. $IC_2$ output at either +8 V or –8 V.
    (c) No oscillations. $IC_2$ output at +8 V. $IC_1$ output at –8 V.

(4) Open circuit connection to $IC_1$, inverting input.

### EXERCISE 8.10 (page 117)

(1) Gate B output stuck at 1.

(2) Input to gate A will assume logic 0. Therefore the output will be 1·2 kHz square waves.

(3) Gate A output stuck at 0.

(4) (a) Output at logic 1 under all conditions.
    (b) 2·4 kHz output only available.

(5) First bistable Q output stuck at 1.

### EXERCISE 8.11 (page 118)

(1) The PIA.

(2) The CPU would be put into a permanent HALT state.

(3) $A800.

(4) CS1 on the PIA would be always enabled irrespective of the state of address line $A_{12}$. The PIA will still be located at address $4800 but additional overwrite to address $5803 would be present.

(5) NMI is non-maskable. It is intended for catastrophic events. IRQ is the standard interrupt request for peripheral devices. It can be masked (i.e. prevented) by setting the I bit in the conditions code register.

(6) A failure in the components of the "max

speed" indicator. To check: apply a logic high (+5 V) via a 2K7 resistor to the base of $Tr_2$. The LED should come on. If it doesn't, isolate the fault further by applying a short between the collector and emitter of $Tr_2$.

(7) System software will only run correctly at power on — not when the reset switch is operated.

(8) $S_1$ open circuit, causing bit 0 (the LSB) to be permanently high.

(9) The 4 bit DAC or the IC power amplifier.

(10) (a) The program will "stick" at the read of $S_5$. The motor will run at full speed.

(b) The "ready to read" indicator will be on continuously.

(c) The address decoder will be off with all of its outputs high. No software operation is possible.

(d) The input pin of the PIA will "float" high allowing a read of the PIA but not a write operation. There will be no drive to the motor or to the indicators.

*Chapter 9*

## EXERCISE 9.2 DAC board (page 132)

(2) (a) $V_{ref}$ will be zero volts and therefore both outputs will also be stuck at zero

(b) Both outputs at zero volts

(c) $IC_2$ acts as a follower with unity gain, thus both outputs will be restricted to a maximum of 2·5 Volts. With the input set to 10000000 the outputs will be only 1·25 V

(d) $V_{out}$ stuck at about +8 V
$\overline{V_{out}}$ stuck low at about −8 V

(e) The voltage gain of $IC_2$ will rise to just over 2, typically 2·2, and it will not be possible to set the maximum output of the circuit correctly.

(f) No effect to $V_{out}$, but $\overline{V_{out}}$ will be incorrect and may stay positive.

(g) With the test input set to 10000000 the resulting output will be in error and read 3·75 V.

## EXERCISE 9.3 (page 134)

(2) (a) Output of $IC_{1b}$
(b) 3
(c) $IC_3$ or $IC_5$

(3) (a) Negative going output pulses only
(b) Positive pulses only
(c) Output restricted to negative output
(d) Output frequency rises to 100 Hz
(e) $IC_4$ output stuck at −8 V

## EXERCISE 9.4 (page 136)

(1) (a) Just greater than 1·25 V
(b) Just greater than 400 mV
(c) 2·5 V or above

(2) Bit 4 of the ADC output stuck high or a collector emitter on the transistor driving bit 4 LED

(3) (a) The clock will run at a high frequency determined by stray and circuit capacitance.

(b) There will be no analogue input and therefore all LEDs will indicate zero.

(c) The OE line will float high allowing the output LEDs to be seen changing state during the successive approximation cycle.

(d) Bit 5 of the output will always remain indicating 0.

(e) Bit 2 of the output will be stuck at 1 under all conditions.

(f) The green LED will not indicate the EOC condition, otherwise there is no effect on circuit operation.

## EXERCISE 9.5 (page 139)

(1) Add a switch or shorting link from the junction of $R_{11}$ and $R_{12}$ to 0 V.

(2) The output level of $IC_1$ will be negative forcing the main timer to be on continuously and the display to indicate zero.

(3) (a) $IC_1$ output high positive. Display at maximum. No timer operation.

(b) Motor on at all times even when the display indicates the temperature as over range.

(c) No reset for $IC_4$ therefore the astable free runs.

(d) The 'on' time of the astable will increase by a factor of three.

(e) The astable will be off under all conditions.

(f) One LED in the bar graph will fail to indicate.

(g) All LEDs in the bar graph display will be off. The rest of the system will function correctly.

# Index